DETAIL Practice

Barrier-Free Design

Principles
Planning
Examples

Oliver Heiss
Christine Degenhart
Johann Ebe

Birkhäuser

Edition Detail

This book is the result of a collaboration between the authors, the Institut für internationale Architektur-Dokumentation GmbH & Co. KG and the advice centre for building without barriers at the Bavaria Chamber of Architects.
The use of the contents of the three German brochures dealing with barrier-free design and construction – "Barrier-free building 1: Barrier-free dwellings" (1992), "Barrier-free building 2: Buildings accessible to the public" (1999), and "Barrier-free building 3: Streets, parks, public circulation/recreation areas and playgrounds" (2001) – is by courtesy of the publishers, the Senior Building Authority in the Bavarian Ministry of the Interior, the Bavarian Ministry for Employment & Social Structure, Families' & Women's Affairs, and the Bavaria Chamber of Architects as well as the illustrators Dipl.-Ing. Michaela Haberkorn, Prof. Dipl.-Ing. Florian Burgstaller and Dipl.-Ing. Andreas Ehrmann.

Authors:
Oliver Heiss, Architect, Urban Planner
Christine Degenhart, Dipl.-Ing. (FH) Architect
Johann Ebe, Prof. Dipl.-Ing. Architect, Urban Planner

Editor:

Project Management:
Steffi Lenzen, Dipl.-Ing. Architect

Editorial services:
Nicola Kollmann, Dipl.-Ing. (FH) Architect

Editorial assistants:
Katinka Johanning, MA; Verena Schmidt;
Melanie Weber, Dipl.-Ing. Architect

Drawings:
Dejanira Bitterer, Dipl.-Ing.; Michael Folkmer, Dipl.-Ing. (FH);
Nicola Kollmann, Dipl.-Ing. (FH) Architect

Translators (German/English):
Gerd H. Söffker, Philip Thrift, Hannover

© 2010 Institut für internationale
Architektur-Dokumentation GmbH & Co. KG, Munich
An Edition DETAIL book

ISBN: 978-3-0346-0577-9

Printed on acid-free paper made from cellulose bleached without the use of chlorine.

This work is protected by copyright. All rights are reserved, specifically the right of translation, reprinting, citation, re-use of illustrations and tables, broadcasting, reproduction on microfilm or in other ways, and storage of the material, in whole or in part, in databases. For any kind of use, permission of the copyright owner must be obtained.

Typesetting & production:
Simone Soesters

Printed by:
Firmengruppe APPL, aprinta druck, Wemding
1st edition, 2010

This book is also available in a German language edition
(ISBN 978-3-920034-27-0).

Bibliographic information published by Die Deutsche Bibliothek.
Die Deutsche Bibliothek lists this publication in the Deutsche Nationalbibliographie; detailed bibliographic data is available on the internet at http://dnb.ddb.de.

Institut für internationale
Architektur-Dokumentation GmbH & Co. KG
Hackerbrücke 6, 80335 Munich
Tel: +49 89 381620-0
Fax: +49 89 398670
www.detail.de

Distribution partner:
Birkhäuser GmbH
PO Box 133, 4010 Basel, Switzerland
Tel: +41.61.568 98 01
Fax: +41.61.568 98 99
e-mail: sales@birkhauser.ch
www.birkhauser.ch

DETAIL Practice
Barrier-Free Design

Contents

7	*Introduction*
9	*On the history of barrier-free design and construction*
10	Historical review
17	*Regulatory principles*
17	Terminology
17	Standardisation
29	*Developments in society*
29	Worldwide demographic developments
30	Demographic change in Germany
32	Consequences for developments in urban planning and the built environment
35	*Design*
35	Limitations and the ensuing demands on the designer
39	Controls and handles
41	Fire protection without barriers
42	Streets, paths, open spaces
50	Stepless accessibility, entrances and doors
52	Vertical access: stairs, ramps, lifts
55	Buildings and workplaces with public access
63	Housing
69	Possible exceptions
75	*Typology*
75	"The third teacher"
80	Housing
86	Sheltered housing and life as an old person
91	Public spaces
93	Public buildings and places of assembly
95	Obtaining provisions, shopping
96	Offices and workplaces
98	*Examples*
98	"Résidence de la Rive" nursing home, Onex
101	Institute for blind children, Regensburg
104	"City Lounge", St. Gallen
107	*Appendix*
107	Glossary
108	Standards, directives, statutory instruments
109	Bibliography
109	Manufacturers, companies and trade associations
110	Index
112	Picture credits

Introduction

We live in a time in which the human race is facing one of its greatest ever challenges. Together, we must tackle the question of sustainability, a term that is repeatedly defined by way of three components: ecology, economy and sociocultural factors. It is precisely this third component that either hinders or allows equal opportunities for all members of our society.

Consequently, "barrier-free" must be understood in an all-embracing sense, for the planning and shaping of our environment without barriers is not just an issue for fringe groups or minorities. Instead, it must be seen as a fundamental task relevant to the whole of society, a task that calls for a long-term and sustainable perspective.

The fact that social circumstances can be deduced from the form of the planned and built environment is something that is frequently only realised afterwards. In order to support integrative equal opportunities, the progression from "building for the disabled" via "building without barriers" to "universal design" is unquestionably desirable. The implementation of this, however, turns out to be complicated because it involves an unspecified number of participants, i.e. "everybody". Therefore, the more imprecisely the requirements are defined, the broader the group of interests and requirements to be integrated, the more likely it will be that essential content will be contradictory. Universal planning and design thus demand that all participants are very willing to compromise.

The draft of the first two parts of the new DIN 18040 "Construction of accessible buildings – Design principles" responds to these necessities in such a way that it does not operate with absolute figures, dimensions and specifications, but instead describes the aims and merely demonstrates typical solutions. Such an approach in a standard is unconventional and leads to an increased need for harmonisation. This book is based on the standards valid at the time of going to press, but does refer to amended or supplemented recommendations and specifications in the new standard where applicable.

This book is divided into three chapters: The first chapter presents the historical side of the development of barrier-free building as well as the sociological, terminological and legislative principles. The benefits of this are that, on the one hand, background information is available for individual design work and, on the other, the facts and figures presented can provide support, especially for liaising with the other members of the planning team, clients, etc., during the project development and draft design stages.
The second chapter, "Planning", has been conceived as an aid for the planning phases. This is where the reader will find definitions of specific requirements in descriptive and graphical form.
The examples presented in the third chapter, "Typology", are intended to inspire the reader and at the same time serve as references that demonstrate potential solutions.

The year in which the German edition of this book was published was special in two ways: in 2009 the "UN Convention on the Rights of Persons with Disabilities" came into force in Germany, and in February DIN 18040 "Construction of accessible buildings – Design principles" was published. The Bavaria Chamber of Architects' Advisory Board on Barrier-Free Building, which was first set up in 1984 through a collaboration with the Bavarian Ministry for Work & Social Affairs, Families & Women's Issues plus the Supreme Building Authority in the Bavarian Ministry of the Interior, provides advice in more than 1000 cases every year.

So this addition to the DETAIL Practice Series, which has been produced with the assistance of the Advisory Board on Barrier-Free Building, attempts to explain not whether, but rather how a barrier-free built environment can have a lasting effect on all our lives. It should help to make building without barriers, without changes of level, an integral, matter-of-course component in design and construction processes and allow the ideas to be turned into reality.

Dipl.-Ing. Lutz Heese
President of the Bavaria Chamber of Architects

On the history of barrier-free design and construction

"And what's your name?"
"Wait, it's on the tip of my tongue."

That is how it all began.
I felt as if I had awoken from a long sleep, and yet I was still suspended in a milky grey. Or else I was not awake, but dreaming. It was a strange dream, void of images, crowded with sounds. As if I could not see, but could hear voices that were telling me what I should have been seeing. And they were telling me that I could not see anything yet, only a haziness ...
A thick, opaque fog which enveloped the noises and called up shapeless phantoms. Finally, I came to a vast chasm and could see a colossal figure, wrapped in a shroud, its face the immaculate whiteness of snow. My name is Arthur Gordon Pym ...
I could hear voices: "Strictly speaking, Signora, it isn't a coma ... No, don't think about flat encephalograms, for heaven's sake ... There's reactivity ..." [1]

That is the opening of Umberto Eco's moving novel *The Mysterious Flame of Queen Loana*. It tells the story of a man who loses his memory and sets off to find traces of his past life in the daily turmoil of the 20th century. Reading these lines, it quickly becomes clear just how fragile and how insecure apparently stable everyday situations really are and how abruptly they can change.

Both the social and the subjective perception and the associated will to integrate have changed. A "disability" is no longer merely an individual, comparatively severe and long-term physical or mental constraint, but rather, a social and socio-cultural one.

The "International classification of impairments, activities and participation" (ICIDH-2), published by the World Health Organisation (WHO) in 1999 and since superseded by the "International Classification of Functioning, Disability and Health (ICF)" (2001), contains an evaluation of impairments, activity limitations and participation restrictions. According to this document it is not a person's inabilities that are critical, but rather their abilities that are relevant for their participation in society. [2]
The integration expert Alfred Sander has formulated the following definition that takes account of the environment: "A disability in a person means an impairment or inability that prevents full integration into his or her complex human-environment system." [3]

Disabilities can be medically diagnosed and categorised as follows:
• Motoric impairments (restricted movement, strength, dexterity and coordination abilities)
• Mental impairments
• Sensorial impairments (visual, hearing, lack of sense of smell or taste)
• Cognitive impairments (speech, learning or mental functions)

The causes are manifold:
• Hereditary impairments (congenital or prenatal effects)
• Impairments suffered shortly before, during or after birth (diseases, physical injuries or ageing processes) [4]

It is frequently the case that such impairments do not occur singly, but rather as a combination of several disabilities of varying seriousness. Those affected must rely on the help of their direct social environment.

People live in social systems and alliances. One of the prime achievements of civilisation was and is caring for the weaker members of society and the development of corresponding welfare mechanisms. Consequently, the organisation and con-

On the history of barrier-free design and construction
Historical review

1 Beguinage, Amsterdam (NL)
  "Fuggerei", Augsburg (D), 1523, Thomas Krebs, endowed by Jacob Fugger the Rich
2 Street
3 Location plan
4 There were initially 52 terrace houses in the "Fuggerei". Every house has one apartment on the ground floor and one on the upper floor.
  The plan layout comprising a living room, kitchen, hallway and two bedrooms is not only typical of the "Fuggerei" – countless farmhouses in southern Germany built between the 15th and 20th centuries have exactly the same internal layout.
  a Kitchen
  b Bedroom
  c Living room
  d Bedroom

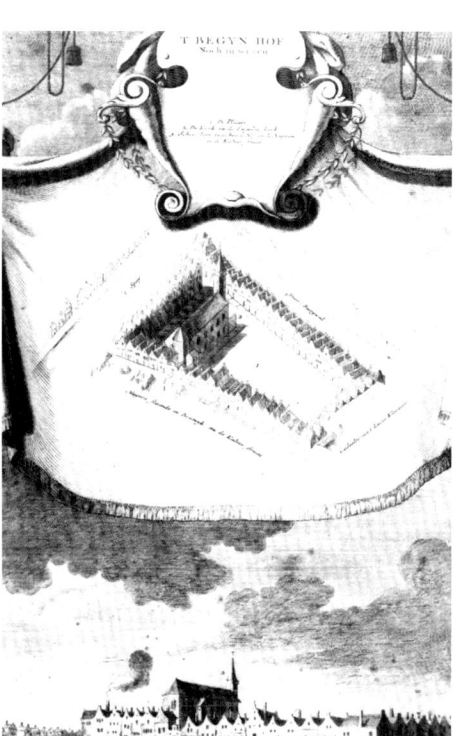

figuration of an appropriate environment of benefit to all has for a long time been a declared objective and is not an invention of the modern age.

Barrier-free design and construction therefore means creating an environment that can be used, preferably independently, by as many members of society as possible, irrespective of their age or physical constitution.

Historical review
The following historical survey is intended to show the beginnings of barrier-free building. Owing to the overall understanding outlined above, these origins cannot be restricted to just "building for the disabled". Social responsibility and its genesis are much more far-reaching. We must assume that welfare facilities essentially originated for three different reasons:
• Care of the sick
• Care of the elderly
• Care of those in need until they are able to care for themselves

Human beings have a social conscience and even in ancient times, e.g. in Persia, this led to the poor and the sick being cared for. In Egypt the temples were also used for treating the infirm. But it is in Sri Lanka and India that we find the first independent establishments designed for the care of the sick.
The parable of the good Samaritan made caring for the sick an obligation for every Christian. In the year 817 the Synod of Aachen declared that every monastery or collegiate foundation should include a "hospital". Christian-run hostels for pilgrims and poor houses were known by this name – its original meaning.

Middle Ages
In the Middle Ages *hospitium* was the name given to a charitable hostel, run by a church or monastery, for pilgrims, the needy, the poor or the sick, or as a refuge ensuring humanitarian aid and welfare needs. Following a decree by Pope Clement V in 1312, hospitals no longer necessarily had to be under the control of the Church. European society prior to 1300 already had well-equipped universities at which theology and philosophy played a great role. However, the causes of illnesses were not understood, nor were there any suitable countermeasures and, unsurprisingly, hygienic conditions were non-existent. Epidemics were unavoidable and, as a result, the plague started to spread in 1349.

The Beghards and Beguines were members of lay Christian communities that started to appear in the 13th century. Irrespective of their wealth or social position, these communities relinquished their personal property to live in semi-monastic, autonomous communities. They dedicated themselves not only to ethical and religious issues, but also practical social work. They looked after the sick, cared for the abandoned, saved the "fallen" and brought up children. The autonomous structure of these communities took on an architectural form in the shape of the so-called Beguinages, especially in the low countries (Fig. 1).
A Beguinage consisted of a cluster of small houses, a chapel and an assembly hall, all grouped around a central courtyard used for growing fruit and vegetables. Even today, this arrangement with comparatively spacious, open urban, common areas appears to be characterised by astonishing neighbourly respect, appropriateness and timelessness.

By comparison, the towns of the Middle Ages were characterised by extremely small forms and dense structures within the existing fortifications. Consequently, within these towns and in monasteries too, planned open spaces were created for

On the history of barrier-free design and construction
Historical review

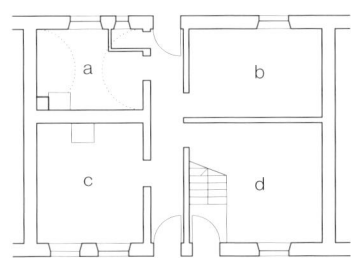

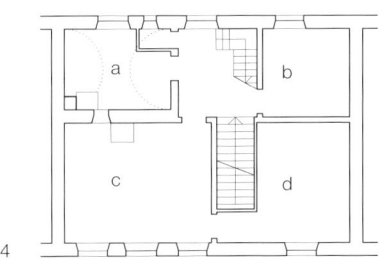

the first time, in the knowledge of their benefit for the community spirit, their restful, identity-building, healthy, sometimes therapeutic effects. Today, hospitals and care facilities normally include gardens for their patients for these very reasons.

The idea of freedom from barriers in the sense of the avoidance of changes of level on a wide scale initially appeared in the urban environment.
With their urban structure and social intentions, the organisational forms of the Beguines can be regarded as the prototype for the "Fuggerei", which was set up in Augsburg in 1521 by Jacob Fugger the Rich. This, a settlement for Augsburg's poor and needy citizens, became the world's first publicly funded housing development. Even today, the annual rent (excluding heating) for an apartment in the "Fuggerei" costs just the nominal equivalent of one Rhenish guilder, currently EUR 0.88, plus three prayers for the founder and his family every day.
The "Fuggerei" is a "town within a town", with its own church, enclosing walls and three gateways. This publicly assisted housing is still financed almost exclusively from the endowment capital (Figs. 2–4).

Renaissance
Jacob Fugger the Rich therefore represents a symptomatic signal for the start of the Renaissance. This was a turning point in history at which the unshakeable divine order underwent a fundamental reform. The ideal of the cultured and at the same time politically active person, the well-educated individual acting responsibly, was formulated.

Foundling hospitals and orphanages
The existence of foundling hospitals or houses, later called orphanages, for infants and small children had been known in central and western Europe since the 9th century. Whereas in ancient times abandoned children were considered to have no rights and were frequently enslaved, Christianity considered such foundlings to be particularly in need of protection. Filippo Brunelleschi expressed this newly found consciousness in the form of his home for abandoned children, the Ospedale degli Innocenti ("hospital of the innocents"), in Florence in 1419. A small revolving wooden cylinder enabled children to be handed in anonymously. The great social achievement here was the fact that these children were not only provided with the essentials, but education and training opportunities as well.

Every person is precious: every child is a personality: every life unique. What makes the Ospedale degli Innocenti so important is not the fact that a solution had been found for saving children from certain death, but that this was attempted despite the forlornness of such a venture (p. 12, Fig. 5). [5]

On the history of barrier-free design and construction
Historical review

5   6

Absolutism
The Reformation and the Thirty Years' War made many people destitute and homeless. Up until this time, homelessness was characterised by charitable work. Protestantism and mercantilism gave rise to social morals that were founded, in particular, on honour, work, material wealth and a personal contribution to the funding of the state. Society, characterised by its hierarchy of different classes, increasingly regarded the poor without gainful employment as a plague that had to be re-educated. Workhouses were introduced in which vagabonds were forced to work for their own betterment. Increasingly, establishments were set up to deal with a whole series of social issues simultaneously, a fact that is revealed in some of their names, e.g. the poorhouse-cum-orphanage-cum-workhouse founded in 1677 in Braunschweig.

Enlightenment
The realisation of the fundamental values and ideas of the Enlightenment as the objectives of the French Revolution – human rights in particular – were among the reasons for profoundly powerful political and socio-political changes across Europe and beyond.

At the start of the 19th century, Napoleon Bonaparte's dream was to turn Paris into a grandiose city of palaces and monuments. At the same time, he decreed that sewerage systems be modernised and pavements, better water supplies, new markets and abattoirs be provided. The rigorous reconstruction of middle-class Paris by Baron Haussmann enabled the weaker members of society to be banished to the suburbs once again. Stephan Lanz has called the rebuilding of Paris a gigantic state gentrification programme, which was also intended to help control insurgency. [6] While the boulevards that had sliced through workers' districts enabled the male middle classes to display their wealth, the poor, the workers and women, who had been consigned to the home, disappeared from public life. This was the period in which the roles of men and women became entrenched. They have remained entrenched to this day.

The discovery of childhood as a phase of life
Our modern understanding of childhood as one phase in life is relatively new. A process known as the "discovery of childhood" first asserted itself in the 17th/18th centuries.
Even today, Johann Heinrich Pestalozzi's holistic approaches are still regarded as fundamental to pedagogics. Elementary education should provide people with a foundation so that they can help themselves and develop the powers that are their natural inheritance.
The first kindergarten, set up in Buda, Hungary, in 1828, was a result of the friendship between Pestalozzi and the Duchess of Brunswick.
The findings of Maria Montessori brought further development to Pestalozzi's pedagogy. Her theory of open teaching can be reduced to the sentence: "Help me to help myself!" Montessori devised special aids that she called "sensorial materials".
In the Casa dei Bambini founded in San Lorenzo in 1907, neglected children from the lower classes were taught according to her pedagogic principles and successfully learned arithmetic and writing within a very short period of time. This all took place in a comparatively good environment, with furniture sizes to suit the children.

Martin Luther had already called for the founding and upkeep of Christian schools in 1524, but it was not until 1717 that King Friedrich Wilhelm I introduced compulsory

On the history of barrier-free design and construction
Historical review

7

schooling for the whole of Prussia, and in Catholic Bavaria it was not until 1802 that it became law to send children to school at the age of six.

Industrialisation
Industrialisation and the associated population explosion in the towns and cities was accompanied by a rapid rise in space requirements, one outcome of which was dense housing structures. Although such structures covered the housing requirements, they were hardly hygienic.

The coming of the railborne infrastructure networks in the early 19th century enabled the frontiers to travel to be overcome – both within and between continents, and the development of vehicles on an industrial scale in the early 20th century, accompanied by the expansion of road networks, made individual mobility possible in ways that had never been seen before.

The age of industrialisation witnessed an intensification of the "removal" of the infirm and the weak, i.e. those who could not take part in the production process. People were assessed purely according to their economic benefits: women were important as industrial workers; but those with disabilities, those who merely disrupted the production process, were rejected.

So those with disabilities had no place in this society. Institutions for the physically and mentally "handicapped" were set up in increasing numbers; their inmates were excluded from society to an unprecedented extent. Separating living and working places resulted in further segregation for the infirm and the disabled. These groups of the population could no longer be cared for within a family unit because now the father and the mother were mostly away from home, working in the factories.

Extended family living forms
Industrialisation brought harsh changes to the structures of rural areas as well. But even though mechanisation had a massive effect on the operation of farms, the once autonomous businesses did not abolish one social accomplishment: the *Ausgedinge* (Fig. 6).
This ensured that once a farmer had passed on his farm to his children, he could be assured of a right to live on the farm, nursing during illness, food, clothes, heating, possibly a monthly pension, too. Depending on the status of the farm, the *Ausgedinge* was either a room, part of a building or a separate, albeit plain, building.

Due to the changing requirements and demands with respect to work, mobility and education, the processes of family establishment have continued to change to the present day. The elderly being cared for directly by their offspring is an exception in the industrialised world. Constraints forced upon us by old age can lead to a self-determined life in familiar surroundings becoming impossible without help. Where this help cannot be provided by other members of the family, various assistance and care models can now be implemented.

Independent households can continue to exist in organisational forms such as self-contained dwellings with ambulatory care, sheltered housing or in flatlets. The options here are continually on the increase, enabling individual requirements and opportunities to be increasingly taken into account.
• Individual sheltered housing (a permanent warden in a housing complex who provides help, with any care services required in ambulatory form)
• Integrated sheltered housing (care centre within housing complex, which also provides daily care services if required)
• Sheltered housing attached to residential home (housing complex located in direct proximity to in-patient care facilities)
• More intensive care services (residential home, home for the elderly, nursing home). [7]

Where more intensive care services are required, residential homes, homes for the elderly and nursing homes can provide the necessary services, but do represent an intervention in the familiar living environment. "Any change to the environment demands a considerable ability to adapt because unfamiliar surroundings call for new lifestyles. Every change is therefore a burden whose distinct, quantifiable intensity increases with the scope of the change." [8]

5 Ospedale degli Innocenti ("hospital of the innocents"), Florence (I), 1419, Filippo Brunelleschi. The opening containing the revolving wooden cylinder (*ruota*) so that children could be handed in anonymously.
6 Modern extended family living: barrier-free *Ausgedinge* in Hof (D), 2006, Seeger & Ullmann
7 City childcare centre, Munich (D), 2006, Atelier SV. The furniture sizes and the wash-basin height are of course just right for the children. Conquering the world of grown-ups begins with the door handle and the riser height of a staircase (the latter should have a second handrail at a lower height for children).

13

On the history of barrier-free design and construction
Historical review

8

In residential homes community living is given preference. Care services are used/offered to a limited extent only. In homes for the elderly, the occupants do not run their own households any longer, but their care needs are relatively minimal. A self-determined lifestyle is given preference here. In nursing homes the priority is 24/7 in-patient care for people with a distinct need for such care.

In summary, we can say that it would be wrong to assume that one universally applicable solution could suit the lifestyle requirements of all older people. Because of the conflicting requirements (maximum freedom, self-sufficiency and independence on the one hand, personal changes due to the changing health situation on the other), it should be ensured that one housing form can adapt to a large number of different requirements and lifestyles. At best, a diverse range of housing forms in different environments results in maximum choice. [9]

Buildings for healthcare
It was in 1784 that Joseph II founded Vienna General Hospital, which for the first time separated the medically ill from others needing care; this had not been the case in hospitals prior to that date. This development turned hospitals into places for medical diagnosis, therapy, teaching and training instead of welfare centres for the poor or old.

Health spas evolved more or less parallel to this. The medical benefits of hot springs have been known since ancient times. The thermal baths of the Romans or the Turkish baths of Arabic regions remain witnesses to that – even today.
In the 18th century Aachen evolved into the leading fashionable spa. Its "Neue Redoute" theatre, designed by Jakob Couven and completed in 1786, became the direct forerunner of the "Kurhaus", the spa administration building.
As a result, sanatoria appeared, the idea of which was to promote the healing process. These experienced their heyday in the late 19th century.

Although Prussia's poor laws of 1891 contained provisions for the treatment and care of the "mentally sick, cretins, epileptics, deaf mutes and the blind who need care", they excluded physically handicapped men, women and children. This latter group had no legal rights to education, training for a profession, or medical treatment. Their financial needs had to be met by their relations, donations and sometimes by aid from poor relief funds. In 1900 there were 13 "homes for the totally crippled" in Germany, run by the Protestant church. These attempted to provide a single form of rehabilitation that covered four related areas: school, orthopaedic treatment, job training and workshops. The focus of "welfare for cripples" was children and youths who were capable of being educated and trained. They were given medical treatment in the homes, educated, trained for jobs and steered towards Protestantism.

Only after the first survey of all physically impaired children in Prussia in 1906/07, initiated by the orthopaedic surgeon Konrad Biesalski, was "welfare for cripples" taken under the national political wing. Up until then the state and the public had been happy to allow the various religious orders to deal with this area. [10]

The International Committee of the Red Cross was set up in 1863 at the instigation of the Geneva businessman Henry Dunant. The aim of the Red Cross is to protect the lives, health and dignity of people, independent of state intervention, and on the basis of voluntary aid. In addition, the

On the history of barrier-free design and construction
Historical review

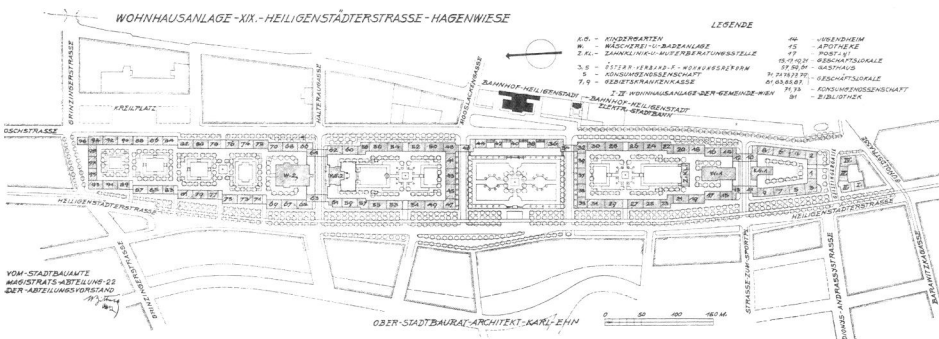

9

Red Cross tries to alleviate the suffering of people in distress irrespective of their nationality or ethnic background, or the religious, ideological or political views of those in need or those administering aid. Just how important this aid organisation is, is shown by their dedication in the many armed conflicts that have taken place since 1863. The first modern artificial limbs, which permitted simple movements, were developed for the many maimed victims of these wars. Regretfully, it is often the severe effects of war that have led to developments in conjunction with barrier-free building.

In 1918 some 1.5 million disabled ex-servicemen had to be re-integrated into the economy of the Weimar Republic. The Centre for Military and Civil Prisoners set up a scheme for the "repatriation of prisoners of war" – welfare and advice centres for prisoners of war who had been released. The main objective of these centres, in addition to material assistance, was to provide information on the legal and new political situations, and offer support during the difficult return to normal life, if indeed this was physically possible at all.

Homes were established for those injured in the war, where the sizes of the rooms allowed movement similar to that in a hospital. The crucial difference, however, was that those confined to wheelchairs had to rely on someone else for their mobility; self-propelled wheelchairs had yet to be invented. It was not until the mid-1960s that wheelchairs started to be adapted to suit their users, so it is quite logical to see why the built environment did not meet the needs of the disabled before this.

Buildings for subsistence welfare
A fundamental answer to the issues of overpopulated structures, poor health conditions and lack of open spaces – all the result of industrialisation – was the proclamation of the motto "air, light and sun" in the 1920s.

Numerous publicly-assisted housing projects were realised in Austrian towns and cities in particular, especially in the 1920s and early 1930s and from the 1950s to the 1970s. In 1923 the social democratic government of "Red Vienna" passed a resolution to build 25,000 new apartments to create affordable rented accommodation with hygienic conditions for the less affluent members of society. [11]
These projects, which were based on the structures of the Beguinages (see pp. 10–11) and the garden cities, were frequently grouped around a central, common courtyard. Varying plan layouts included common amenities and services, e.g. childcare.

Even today, Vienna's housing associations organised on a cooperative basis, which are generally non-profit-making, cover about 80 % of the city's housing needs. In Germany the share of publicly assisted housing is currently only about 20 %, which, on the one hand, results in totally different ownership issues but, on the other, totally different living costs (e.g. rents).

Building for the disabled – building without barriers – universal design
Up until the 1960s disabilities in political and scientific discourse were defined exclusively as physical, mental or intellectual impairments of individuals that prevented them from participating in the activities, accomplishments and mobility of society and excluded them from productive gainful employment.
Consequently, policies for improving conditions for the disabled were aimed at adapting the physical, sensorial and mental functions of those affected to normal socio-cultural expectations. It seemed that the social integration problem could be solved only if sufficient material social services and welfare systems could be created, successful rehabilitation measures and medical therapies guaranteed.

It was not until the 1970s that impairments were no longer seen as an exclusively individual problem, but rather a society-related one. In his governmental statement of 1969, Federal Chancellor Willy Brandt called for the quality of a democratic society to be measured by its policies for the disabled.

8, 9 Some 5000 people live in the 1350 apartments of this approx. 1 km long building complex. Karl-Marx-Hof, Vienna (A), 1930, Karl Ehn

15

On the history of barrier-free design and construction
Historical review

The social exclusion of persons with disabilities became less and less compatible with the social claims of the expanding Federal (i.e. West) German welfare state. The growing number of emancipation, organisational and self-help movements for people with disabilities gradually led to the groups affected participating in the discussions on policies for the disabled. The committees drafting the social-liberal strategies for those with disabilities were aware of the fact that they had to deal not only with the individuals directly affected, but also with their environments and society as a whole, if equal rights and equal opportunities for those with disabilities were to be achieved.

It was at this time that constructional and technical aspects first started to be seen as barriers to mobility and lifestyles. Work on standards for planning and building began in the early 1970s. They tried to formulate definitions for the built environment that take not only the needs of the disabled into account, but those of all people. In the beginning, however, the standardisation requirements focused merely on young, male wheelchair users. [12]

The term "barrier-free" first appeared in the mid-1980s. It reflected the need to focus not only on the individuals, but also on their surroundings.
During the 1990s integration policies started to apply more and more the understanding of the equality principle rather than simply the classical welfare and social services. In Germany, the Anti-Discrimination Act of the mid-1990s, the equality legislation of the federal government and the federal states (2002) and the Sex Discrimination Act of 2006 demonstrated this.

Current and future policies are and will be characterised by the attempt to dismantle barriers, promote self-determination and deal with disabled people as equals. One expression of this development can be seen in the changing definitions: "universal design", "inclusive design" or "accessibility for all" are increasingly replacing the term "barrier-free". [13]
History has shown that such changes to definitions are more than just semantics. They are an expression of real change in society. Such developments have in the meantime helped both those affected and others to see diversity as "normal".

[1] Eco (trans. Brock), 2004, p. 7ff.
[2] WHO, 2001
[3] Eberwein/Knauer, 2002
[4] Rau, 2008, p. 13ff.
[5] Mayer, 2008
[6] Lanz, 2008, p. 295
[7] Marx, 2009, p. 29
[8] Hugues, 1975, p. 21
[9] ibid.
[10] Stadler/Wilken, 2004
[11] Degenhart, 2008, p. 16
[12] Bösl, 2009
[13] Herwig, 2008

Regulatory principles

The following pages are intended to provide an overview of current legislation, directives, standards and legislative levels in Germany. Firstly, a number of key terms are explained.

Terminology
Disability
"Persons are handicapped [disabled] if their physical function, mental ability, or mental health deviate from the condition that is typical for their age for a period that will last longer than six months and, if as a result, their participation in social life is impaired." [1]
The typical disabilities that occur as a direct result of old age are therefore not classed as disabilities in the meaning of the German Social Code (SGB) IX.

Severe disability
Persons are severely disabled if their degree of physical, mental or emotional impairment exceeds 50 %, and not only temporarily. Unfortunately, the degree of disability is frequently equated to a physical or mental performance constraint. Severely disabled people with a degree of disability of 100 % can still achieve full performance in their chosen activity provided their respective disability does not affect the particular task.
So it is always the nature of the constraint in conjunction with the activity that is important. [1]

Barrier-free
"Buildings and other facilities, means of transport, technical artefacts, information dissemination systems, acoustic and visual information sources and communication equipment plus other configured lifestyle areas are considered to be barrier-free when they are accessible to, and usable by, disabled people in the generally normal way without particular difficulties and certainly without any assistance." [2]

A barrier-free environment is especially helpful for the following groups of people:
• Wheelchair users, also those with upper body disablement
• Mobility impaired
• Blind
• Visually impaired
• Deaf
• Hearing impaired
• Persons with other impairments
• Older people
• Children
• Exceptionally short/tall people

Wheelchair accessibility
The following may be required to ensure accessibility for wheelchair users:
• Barrier-free buildings and other facilities
• Means of transport
• Technical artefacts
• Information dissemination systems
• Acoustic and visual information sources and communication equipment
• Other configured lifestyle areas

Universal design
Products, environments, programmes and services that comply with the "universal design" concept can be used by all people to the maximum possible extent without the need for any adaptation or special configuration.

Principle of two senses
Every movement and activity in space presumes the perception of the stimulation of various senses plus their association, interpretation and reception. In order to be able to comprehend our environment, the senses of sight and hearing are extremely important. If a person's perception options cannot be used to the full, then the body normally tries to deal with this deficit by exploiting the remaining options and also by way of compensatory measures. Alternative perceptions according to the principle of two senses are then possible if information is available simultaneously for two of the five senses. This principle is particularly helpful for the choice of materials and when planning orientation or guidance systems (acoustic signal in lift plus, for example, labelling in Braille).

Standardisation
Codes of practice and standardised sizes are extremely helpful for planning activities because they formulate the design conditions exactly and hence define generally acknowledged principles. But they therefore also imply a restriction on the options that can be implemented.

Definitions applicable worldwide
The United Nations' "Convention on the Rights of Persons with Disabilities", which has been law in Germany since 26 May 2009, sends a signal to the world for the strengthening of the rights of the more than 600 million people with disabilities. More than two-thirds of them live in so-called developing countries. For them, the Convention represents the first-ever universally applicable agreement. It stipulates human rights, e.g. the right to life or the right to freedom, taking into account the specific situation of persons with disabilities.

10 Barrier-free ramp linking old and new, Bavaria Chamber of Architects, Munich (D), 2002, realgrün landscape architects, with Drescher & Kubina

Regulatory principles
Standardisation

T1: Technical construction regulations adopted in building legislation

Federal state	DIN 18024-1 "streets, squares, paths …"	DIN 18024-2 "publicly accessible buildings and workplaces"	DIN 18025-1 "dwellings for wheelchair users"	DIN 18025-2 "accessible dwellings"
Baden-Württemberg[1]	–	Adopted	Adopted	Adopted
Bavaria[2]	–	–	–	–
Berlin	Adopted, with the exception of sections 13, 15, 16, 19	Adopted, with the exception of sections 6 sent. 4, 8, 11 sent. 1, 13, 14, 16	Adopted	Adopted
Brandenburg	Adopted, with the exception of sections 8.4, 8.5, 9, 10, 10.1 sent. 2, 12.2, 13, 14, 15, 16, 19	Adopted, with the exception of sections 6 sent. 4, 8, 11 sent. 1, 13, 14, 16	Adopted	Adopted
Bremen	Adopted, with the exception of sections 8.4, 8.5, 9, 10.1 sent. 2, 12.2, 13, 14, 15, 16, 19	Adopted, with the exception of sections 1, 16; special regulations apply for sections 6 sent. 4, 7.1, 7.3 sent. 1, 10, 11 sent. 1 & 2, 13, 14	Adopted	Adopted
Hamburg	–	Adopted, with the exception of sections 1, 6 sent. 4, 11 sent. 1, 13, 16	Adopted, with the exception of sections 1, 5 No. 5.1, 6 No. 6.3, 11 para. 7	Adopted, with the exception of sections 1, 5 No. 5.1, 6 No. 6.3
Hesse	Adopted	Adopted	Adopted	Adopted
Mecklenburg-Western Pomerania	–	Adopted, with the exception of sections 6 sent. 4, 8, 11 sent. 1, 13, 14, 16	–	–
Lower Saxony	–	Adopted, with the exception of sections 1, 11 sent. 1, 14, 16, 18 sent. 1; section 6 sent. 4 may not be used for fire-resistant doors	Adopted, with the exception of sections 1, 5.1, 6.1, 6.3, 6.4, 6.5, 6.6, 10, 11 para. 7	Adopted, with the exception of sections 1, 5.1, 6.1, 6.3, 6.4, 8, 11
North Rhine-Westphalia	–	–	–	–
Rhineland Palatinate	–	Adopted, with the exception of sections 1, 11 sent. 1	Adopted, with the exception of section 1	Adopted, with the exception of section 1
Saarland	Adopted, with the exception of sections 8.4, 8.5, 9, 10.1 sent. 2, 12.2, 13, 14, 15, 16, 19	Adopted, with the exception of sections 6 sent. 4, 8, 11 sent. 1, 13, 14, 16	Adopted	Adopted
Saxony	Adopted, with the exception of sections 8.4, 8.5, 9, 10.1 sent. 2, 12.2, 13, 14, 15, 16, 19	Adopted, with the exception of sections 6 sent. 4, 8, 11 sent. 1, 13 sent. 2–4, 14, 16	Adopted	Adopted
Saxony-Anhalt	Adopted, with the exception of sections 8.4, 8.5, 9, 10.1 sent. 2, 12.2, 13, 14, 15, 16, 19	Adopted, with the exception of sections 6 sent. 4, 8, 11 sent. 1, 13, 14, 16	Adopted	Adopted
Schleswig-Holstein	Adopted	Adopted	Adopted	Adopted
Thuringia	Adopted, with the exception of sections 8.4, 8.5, 9, 10.1 sent. 2, 12.2, 13, 14, 15, 16, 19	Adopted, with the exception of sections 6 sent. 4, 8, 11 sent. 1, 13, 14, 16	Adopted	Adopted

Technical construction regulations that have been adopted are technical codes of practice that have been incorporated into building legislation by the supreme building authorities of the individual federal states by way of public announcements. Only those technical construction regulations that are indispensable for fulfilling the requirements of building legislation are used.
Supplements:
[1] Controls: "The standard dimensions for grip and operating heights is 850 mm (dimension to centre-line) above FFL; deviations in the range 850–1050 mm are permitted if necessary."
[2] Together with the Ministry of Social Affairs and the Bavaria Chamber of Architects, the supreme building authority has, however, published commentaries on DIN 18024 and 18025. The Bavarian Parliament has included the integration of audio induction loops in its equality legislation to improve communications for those with impaired hearing

Regulatory principles
Standardisation

11 Attractive public spaces provide multi-functional – not always completely foreseeable – options that can be used by different social groups. Georg-Freundorfer-Platz, Munich (D), 2002, Levin Monsigny landscape architects

European treaty
The heads of state of the European Union agreed on a constitution at their summit in Brussels on 18 June 2004. This "Charter of Fundamental rights of the European Union" includes definitions regarding non-discrimination and the integration of persons with disabilities – in articles 21 and 26. On 17 March 2008 the European Union adopted a resolution of the Council of the European Union and the representatives of the Governments of the Member States on the situation of persons with disabilities in the European Union (2008/C 75/01). This document recommends that the Member States adopt the UN's "Convention on the Rights of Persons with Disabilities" and embody it in their national legislation.

Standards valid in Germany
Standards, directives and recommendations incorporate the state of the art and may be used by anybody without them necessarily being legally binding in principle. They only become legal documents when they are referred to or adopted in legislation. DIN 18024 and DIN 18025 form part of the building regulations of a number of Germany's federal states. The application of the standards adopted in building authority legislation is covered by the technical construction regulations. However, in cases of conflict, a standard can still be consulted even if it does not form part of the legislation (Tab. T1).

Barrier-free planning concepts must always be considered in the light of different restrictions. The first German standards dealing with barrier-free building were introduced in the mid-1970s: DIN 18025 "Dwellings for seriously disabled persons; design principles" part 1 "Dwellings for wheelchair users" and part 2 "Dwellings for blind persons and those having essential difficulty in seeing". Furthermore, part 1 ("Streets, places and ways") and part 2 ("Publicly accessible buildings") of DIN 18024 "Construction measures for disabled persons and elderly human beings in the public field; design principles" were also published in this period. (The titles of these standards have since been amended.) However, sensorial and cognitive impairments were not fully included in the standardised definitions until the publication of disabled persons equality legislation and the changes associated with that, e.g. in building regulations.

DIN 18040 "Construction of accessible buildings – Design principles" part 1 "Publicly accessible buildings" and part 2 "Dwellings" were published in 2009. These are intended to replace DIN 18024 "Construction of accessible buildings" part 2 "Publicly accessible buildings and workplaces, design principles" as well as parts 1 and 2 of DIN 18025 "Accessible dwellings". These are now the critical documents for barrier-free design in Germany. However, the revised definitions in DIN 18040 do not rule out the fact that design and construction procedures to abolish barriers in accordance with the regulations for one group of affected persons can lead to the creation of a barrier for another group. For example, lowering kerbs to make it easier for wheelchair users to cross the road can lead to visually impaired persons losing a clear orientation aid unless the detailing is carried out properly (Tab. T2).

At this point it becomes clear that it is impossible to accomodate the diversity of individual, different requirements in a single, legally binding regulation, despite the availability of a standard covering the anthropometric data of the human body (DIN 33402). The introduction of DIN 18040 will lead to the application of the so-called performance concept. This standard therefore leaves it up to the user to choose suitable means for appropriate design and construction that will satisfy the requirements of the respective aims in each specific instance. Only examples of potential solutions are described in the standard. In the event of a conflict, however, it is up to the designer to prove that the objectives have been fulfilled when the details do not correspond to the examples given in the standard. [3]
This approach could permit the use of more comprehensive integration processes because it contains less restrictive stipulations. However, it also represents greater uncertainties and liability risks for the designer.

Legislation at federal government level in Germany
• Basic Law for the Federal Republic of Germany (*Grundgesetz*, GG)
"No person shall be favoured or disfavoured because of sex, parentage, race, language, homeland and origin, faith, or religious or political opinions. No person shall be disfavoured because of disability."
GG, art. 3, para. 3, sent. 1; Act for Amending the Basic Law, 27.10.1994, Federal Gazette I, 1994, p. 3146

• General Anti-Discrimination Act (*Allgemeines Gleichbehandlungsgesetz*, AGG) This is federal legislation that is intended to eliminate and prevent disadvantages due to reasons of "race, ethnic origin, gender, religion or ideology, a disability, age or sexual identity". AGG, 14 Aug 2006, Federal Gazette I, 2006, p. 1897

• Disabled Persons Equality Act (*Behindertengleichstellungsgesetz*, BGG) This defines the terminology for disabilities in cl. 3 and also barrier-free design and construction in cl. 4.
BGG, 27 Apr 2002, Federal Gazette I,

19

Regulatory principles
Standardisation

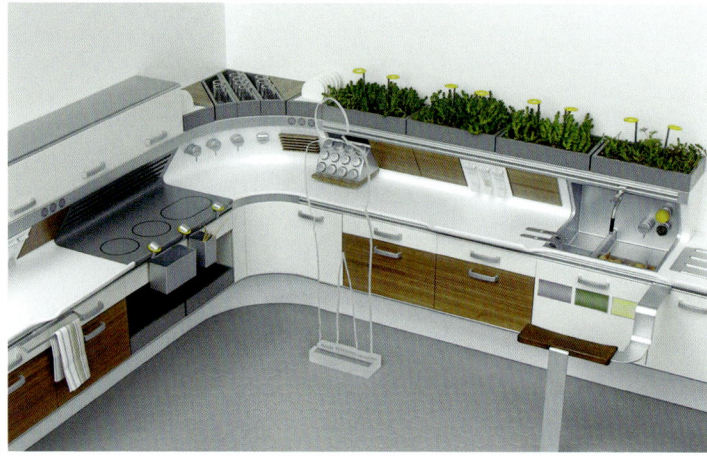

12 The kitchen is a key element in a home that should be used independently for as long as possible. Cooking and dining areas should therefore be given special attention at the design stage, not only for ergonomic reasons because of the working procedures, but also because of their social significance. Kitchen concept "50 Plus", Diana Kraus, diploma thesis, Coburg University of Applied Sciences, 2006

2002, p. 1674 Behinderung. BGG v. 27.04.2002, BGBl. I 2002, S. 1674

- Federal Building Code (*Baugesetzbuch*, BauGB)
  "(6) In the preparation of land-use plans, attention is to be paid in particular to the following:
  … 3. the social and cultural needs of the population, in particular those of families, the young and the elderly and those with handicaps, as well as to the requirements of the education system and the need for sports, leisure and recreational facilities. …"

BauGB, cl. 1, para. 6, "Scope, definition and principles of urban land-use planning"

- Homes Act (*Heimgesetz*, HeimG)
  This act specifies standards for the residential care of older people, those in need of care or disabled adults. It contains provisions for the protection of the residents of homes and covers homes for accommodating people who as a result of their age, disability or care needs must be cared for in a home. HeimG cl. 11 formulates requirements for the operation of a home. HeimG, 7 Aug 1974, last amended on 31 Oct 2006

- Social Code (*Sozialgesetzbuch*, SGB)
  This law stipulates that in public- and private-sector companies with at least 20 members of staff, at least 5 % of the employees should be severely disabled people. On the whole, the SGB contains provisions for disabled persons and those threatened with disablement which guarantees them the right to maximum self-determination and participation in society, provides assistance for their step-by-step integration and encourages self-help. SGB IX – Rehabilitation and participation of disabled persons, 19 Jun 2001, last amended on 31 Dec 2008, cl. 71, para. 1

T2: Selection of standards

Standard	DIN 18024-1	DIN 18024-2	DIN 18025-1	DIN 18025-2
Title	Barrier-free built environment – Part 1: Streets, squares, paths, public transport, recreation areas and playgrounds – Design principles	Construction of accessible buildings – Part 2: Publicly accessible buildings and workplaces, design principles	Accessible dwellings; dwellings for wheelchair users, design principles	Accessible dwellings; design principles
Content	Valid for streets, paths and squares, public circulation areas, parks and gardens, recreational areas and their accesses. Valid for new works and applies similarly to conversions, modernisations and changes of use. The following areas are defined: • Movement areas and passing places • Pedestrian circulation zones • Stairs, ramps, lifts • Parks, gardens, recreational areas • Securing of building sites • Bus/tram stops, railway platforms • Car parking areas • Fittings (orientation, signs, lighting)	Valid for the planning, design and provision of publicly accessible buildings or parts thereof plus workplaces and their external facilities. The standard does not apply to hospitals where enhanced requirements apply in many areas. The following areas are defined: • Movement areas and passing places • Doors • Stepless designs, ramps, stairs • Lifts • Walls, ceilings, floor coverings • Controls • Orientation aids • Sanitary facilities, first-aid rooms • Workplaces, leisure facilities • Places of assembly, sports centres, restaurants • Car parking areas	Valid for the planning, design and provision of new wheelchair-accessible, rented accommodation, cooperative-owned housing and corresponding residential complexes. Applies similarly to roof space and other conversions and modernisations. The following areas are defined: • Areas • Doors • Stepless designs, ramps • Lifts • Stairs • Kitchens • Bathrooms • Car parking areas • Walls, windows • Technical equipment	Valid for the planning, design and provision of new barrier-free, rented accommodation, cooperative-owned housing and corresponding residential complexes. Applies similarly to roof space and other conversions and modernisations. The following areas are defined: • Areas • Doors • Stepless designs, ramps • Lifts • Stairs • Kitchens • Bathrooms • Car parking areas • Walls, windows • Technical equipment

Further standards relevant for barrier-free design and construction can be found in the appendix.

Regulatory principles
Standardisation

- Places of Work Act (*Arbeitsstätten-verordnung*, ArbStättV) and Places of Work Directives (*Arbeitsstättenrichtlinien*, ASR)
These formulate requirements regarding the barrier-free design of workplaces plus the associated doors, circulation routes, escape routes, emergency exits, stairs, orientation systems and sanitary facilities.
ArbStättV, in particular cl. 3 (2) "Establishment and operation of workplaces for people with disabilities", 12 Aug 2004, last amended on 18 Dec 2008

- Minimum Building Regulations for Homes (*Heimmindestbauverordnung*, HeimMindBauV)
This legislation prescribes minimum building requirements for homes for the elderly, residential homes and nursing homes for adults.
HeimMindBauV, 27 Jan 1978, last amended on 25 Nov 2003 (Tab. T3)

- Model Building Code (*Musterbauordnung*, MBO)
In Germany building legislation falls under the jurisdiction of the individual federal states. Each state has its own building regulations (Tab. T5). However, the ministries responsible in the federal states meet at regular intervals to discuss which building legislation provisions in all the federal states are sensible from the point of view of greater uniformity. The result of these deliberations is the so-called Model Building Code which, however, is not a legally binding document in the individual federal states. "(1) In buildings with more than two dwellings, the dwellings on one floor must be accessible without barriers. In these dwellings the living rooms and bedrooms, one toilet, one bathroom

Standard	DIN 18030	DIN 18040-1, draft 2009 [1]	DIN 18040-2, draft 2009 [1]	DIN 77800
Title	Barrier-free building	Construction of accessible buildings – Design principles – Part 1: Publicly accessible buildings	Construction of accessible buildings – Design principles – Part 2: Dwellings	Quality requirements for providers of "assisted living for the elderly"
Content	The work on this standard, the task of which was to develop DIN 18024 and 18025 further, has been abandoned after 10 years because a consensus could not be reached.	Valid for the barrier-free planning, design and equipping of publicly accessible buildings and their external facilities required for access to and usage of the building. The standard applies to new works and should be applied similarly for the planning of conversions and modernisations. The following areas are defined: • External access, paths • Car parking areas • Access and entrance zones • Internal access • Doors, corridors • Floor coverings • Lifts, stairs, escalators, ramps • Wheelchair storage areas • Warning, orientation, informing, guiding • Controls, communication systems, equipment • Ticket windows, checkouts, security barriers • Alarms, evacuation • Rooms, venues for events • Sanitary facilities • Changing rooms • Swimming and therapy pools	Valid for the planning, design and equipping of dwellings, also buildings containing dwellings, and their external facilities required for access to and usage of the building, or the residential parts thereof. Also takes into account their unrestricted use by wheelchair users (dimensions suitable for wheelchairs are specifically indicated in the standard). The standard applies to new works and can be applied similarly for the planning of conversions and modernisations. The following areas are defined: • External access, paths • Car parking areas • Access and entrance zones • Internal access • Doors, corridors • Floor coverings • Lifts, stairs, escalators, ramps • Wheelchair storage areas • Warning, orientation, informing, guiding • Controls, communication systems, equipment • Rooms in dwellings • Corridors, doors, windows • Living rooms, bedrooms, kitchens • Sanitary facilities • Open-air seating	The standard contains requirements regarding the transparency of services on offer, the services to be provided, the residence options on offer, the wording of contracts and quality assurance measures. It provides customers, building owners and service providers with a uniform quality standard and therefore acts as an alternative to the multitude of regional and local regulations which are frequently impossible to compare with each other. This overcomes the regionalisation of the certification schemes already evident and the associated ambiguities. The standard contains many requirements regarding the duty to provide information, including: • housing complex • dwellings • basic services • optional services • costs and funding

[1] Part 1 has been approved and is due to be published in the late summer of 2010. [2] Part 2 is due to be approved soon and should also be published in the late summer of 2010.

Regulatory principles
Standardisation

T3: The requirements of the Minimum Building Regulations for Homes (*Heimmindestbauverordnung*, HeimMindBauV)

No.	cl. No. (with cl. 29 if necessary)	Content of provision
General requirements (Nos. 1–21)		
1	2	It must be possible to reach all residential and nursing places from a generally accessible corridor.
2	3 para. 1	Corridors may not include any steps or, if there are any, then suitable ramps must be provided alongside as well.
3	3 para. 2	Corridors to nursing places must permit the transport of bedridden residents.
4	3 para. 3	Corridors and stairs must be provided with permanent, secure handrails on both sides.
5	4	A lift must be provided when more than one storey height is to be accessed or wheelchair users are accommodated in storeys with stepped access.
6	5	Floor coverings in rooms and circulation areas used by residents must be of the non-slip variety.
7	6 para. 1	Light switches must be easy to operate.
8	6 para. 2	Night-time lighting must be used to illuminate staircases and corridors during the hours of darkness.
9	6 para. 3	Power sockets for operating reading lamps must be available in living rooms, bedrooms and common rooms. In bedrooms these sockets must be close to the beds.
10	7	Rooms for accommodating persons requiring nursing services must be equipped with an emergency call system that can be operated from every bed.
11	8	A telephone must be provided in every building so that residents can be reached; it must be possible for non-bedridden residents to use the telephone without being overheard by others.
12	9 para. 1	Living rooms, bedrooms and sanitary rooms must be accessible from outside in emergencies.
13	9 para. 2	Doors to nursing places in nursing homes and nursing departments must be wide enough to permit the transport of bedridden residents.
14	10 para. 1	Baths and showers in common facilities must include privacy screens that can be moved into position when the facilities are in use.
15	10 para. 2	There must be safe means of getting into and out of baths.
16	10 para. 3	Baths, showers and WCs must be fitted with handrails.
17	10 para. 4	An adequate number of suitable sanitary facilities must be available in establishments used by wheelchair users.
18	11	An adequate number of suitably sized utility rooms must be available where such services are not provided by external providers.
19	12	Suitable heating systems that guarantee a temperature adapted to the needs of residents must be provided for all rooms, staircases, corridors and sanitary facilities.
20	13 sent. 1	There should be no steps between public circulation areas and the entrance level of any building used by residents.
21	13 sent. 2	The entrance must be illuminated.
Special requirements for residential homes (Nos. 22–35)		
22	14 para. 1 sent. 1	Residential places for one person must have at least one bedsit room with a floor area of 12 m^2, or 18 m^2 for two persons.
23	14 para. 1 sent. 2 & 3	Residential places for more than two persons require express permission from the authority responsible. Residential places for more than four persons are not permitted. The additional floor area for the third and fourth persons must be at least 6 m^2 each.
24	14 para. 3 sent. 1	Residential places for up to two persons must be fitted with one wash-basin with hot and cold water supplies.
25	14 para. 3 sent. 2	Residential places for more than two persons must be fitted with a second wash-basin with hot and cold water supplies.
	15 para. 1	The following must be provided in every establishment:
26	No. 1	adequate cooking options for the residents,
27	No. 2	a room for storing the residents' possessions,
28	No. 3	where rooms contain several beds, a single room in the meaning of cl. 14 for temporary use by residents,
29	No. 4	a mortuary, if quick removal of bodies cannot be guaranteed.
30	15 para. 2	Where an establishment consists of several buildings, the requirements according to para.1, Nos. 1 and 3 must be satisfied in every building.
31	16 para. 1	The establishment must have at least one common room with a usable floor area of 20 m^2. In establishments with more than 20 residents, a usable floor area of at least 1 m^2 per resident must be available.
32	17	Every establishment must include a room for exercise therapy or gymnastics if there are no suitable gymnastics or therapy facilities available within a reasonable distance of the establishment which can be used regularly by the residents. Common rooms according to cl. 16 may be used for this.
33	18 para. 1	At least one WC with handrinse basin must be available on the same floor for every eight residents.
34	18 para. 2	At least one bath or shower must be available in the same building for every 20 residents.
35	18 para. 3	Baths in common bathrooms in nursing departments must be installed so that they are accessible from both longitudinal sides and one end.

and the kitchen or kitchenette should be accessible for wheelchairs. …"
"(2) In buildings accessible to the public, the parts serving the general visitor circulation must be accessible without barriers to disabled persons, old people and persons with small children, and must be able to be used appropriately without assistance. These requirements apply in particular to
1. cultural and educational facilities,
2. sports and leisure facilities,
3. health facilities,
4. office, governmental and court buildings,
5. retail premises and restaurants,
6. parking places, garages and WC facilities."
"(3) Buildings according to para. 2 must be accessible via a stepless entrance with a clear width of at least 0.90 m. A movement area of adequate size must be available in front of doors. Ramps may be inclined at no more than 6 %; they must be at least 1.20 m wide and have permanent, secure handrails on both sides. There must be a landing at the start and end of each ramp, every 6 m an intermediate landing. Landings must be at least 1.50 m long.
Stairs must be fitted with handrails on both sides which continue across landings and window openings and beyond the last steps. Stairs must be fitted with solid risers.
Corridors must be at least 1.50 m wide. One WC room must also be suitable for and accessible to wheelchair users; it must be indicated."
"(4) Paras. 1 to 3 do not apply where the requirements can only be satisfied with excessive additional costs owing to difficult terrain conditions, owing to the installation of a lift that would otherwise be unnecessary, owing to unfavourable existing constructions or would affect the safety of people with disabilities or old people."
Extract from MBO, cl. 50, "Barrier-free building", paras. 1–4, Nov. 2002

In German-speaking countries, the definitions given in DIN 18040 will result in a unified fixing height for door handles of 850–1050 mm above finished floor level, for example. In Scandinavia and the UK the standardised movement areas are somewhat larger. But they have less restrictive clauses regarding, for example, separate sanitary facilities for men and women. Tactile guidance systems in public areas, however, have long since been standard. On the whole, the integration efforts do seem to be leading to the establishment of pan-European definitions for barrier-free design and construction.

Tab. T4 shows assistance options and corresponding jurisdictions.

13 Grass ramp, Brandhorst Museum, Munich (D), 2008, Adelheid Gräfin Schönborn, with Sauerbruch Hutton

[1] Social Code IX: cl. 2, para. 1, 19 Jun 2001
[2] cl. 4, *Behindertengleichstellungsgesetz*, *Barrierefreiheit*, 2002
[3] Frohnmüller, 2009, pp. 10–11
[4] see also Rau, 2008, p. 316ff.

Regulatory principles
Standardisation

T4: Assistance programmes

Level	Assistance from…	Form of assistance
Federal government	State Employment Agency	• Helps persons with physical, mental and emotional impairments to be integrated into the world of work • Can provide subsidies for constructional modifications in order to establish or maintain a workplace
	State Integration Agency	• SGB IX cl. 101, para. 1, No. 1 • For the procurement, equipping and maintenance of a dwelling suitable for the needs of the disabled
	KfW Bank	Owner-occupied housing programme • First- and second-time purchases with 10-year no-sale bond • Loans of up to 30 % of the total cost, max. EUR 100,000 • Cumulative options with other assistance programmes must be checked
		Conversion to suit the needs of the elderly • Funding up to 100 % of the eligible costs, max. EUR 50,000 per dwelling • Can be combined with energy-efficiency upgrade measures
	Health insurance	• Provides assistance in order to compensate for physical impairments • Can pay for costs of therapy and of mitigating the problems with a dwelling and its interior furnishings and fittings
	Statutory nursing care insurance	• (Nursing care insurers) can provide subsidies for persons acknowledged as needing care in care categories 1–3 • Can provide assistance for adapting a dwelling
	Rehabilitation organizations	• May assist with financing for the procurement, equipping and upkeep of a dwelling suitable for the needs of the disabled • Information available via the Federal State Pensions Regulatory Authority or employers' liability insurance association
	Pensions Regulatory Authority	• Discretionary payments • Assists with conversion work or home care services merely within the scope of re-integration into working life
	Social Security Administration	Restoration of self-sufficiency • Can provide older and disabled persons with help to improve their lifestyles when the conditions based on the Federal Social Security Act (*Bundessozialhilfegesetz*, BSHG) are fulfilled The legal basis in detail: • Integration assistance (BSHG cl. 39, 40, 49) in order to eliminate a disablement and its consequences and enable maximum self-sufficiency for the person affected • Care services (BSHG cl. 68, 69) in order to provide persons needing care with the appropriate services and aids • Help for the aged (BSHG cl. 75), also for procuring and maintaining a dwelling that meets the needs of older people
	Pensions & Relief Office	The basis for services are: • Federal War Victims Relief Act (*Bundesversorgungsgesetz*, BVG), SGB I cl. 68 • Crime Victims Compensation Act (*Opferentschädigungsgesetz*) • Severely Disabled Persons Act (*Schwerbehindertengesetz*) • War Victims Welfare Act (*Kriegsopferfürsorgegesetz*) • Act for Aid for Political Prisoners Outside the FRG (*Häftlingshilfegesetz*) • Federal Communicable Diseases Act (*Bundesseuchengesetz*) • Severely disabled persons can receive financial assistance for equipping or modifying an existing dwelling (BVG cl. 27c) • Assistance with the procurement of a dwelling suitable for the needs of the disabled
Federal state governments		

Baden-Württemberg	Landeskreditanstalt	• Assists with the construction, purchase or modification of dwellings for disabled persons within the scope of
		  - standard assistance
		  - additional assistance

Bavaria	Landesboden-kreditanstalt	• Assists with owner-occupied (new) dwellings on the basis of the Bavarian Housing Assistance Act (*Bayerisches Wohnraumförderungsgesetz*, BayWoFG) • Loans, plus grants of EUR 1500 per child • Additional grant of EUR 10,000 for disabled persons
		• Assists with rented accommodation in multi-occupancy buildings on the basis of the BayWoFG and the housing assistance provisions of 2008 • Supplementary assistance options for disabled persons
		• Modification of dwellings (rented and owner-occupied) to suit a disability (conversion) on the basis of the BayWoFG • Non-repayable loans of up to EUR 10,000
		• Bavarian interest-reduction scheme for new construction, first- and second-time purchases with 10-year no-sale bond • Loans of up to 30 % of the total cost, max. EUR 100,000
		• Modernisation of rented accommodation in multi-occupancy buildings plus nursing places in residential homes for senior citizens • 30-year repayment time, 10 years of reduced interest • Energy-efficiency upgrades for dwellings and conversions to suit the needs of the elderly
		• Replacement of residential homes for senior citizens in Bavaria • Capital market loans • Homes for disabled persons

Regulatory principles
Standardisation

Level	Assistance from…	Form of assistance
Bavaria (contd.)	Local governments in Munich, Augsburg & Nuremberg	• New construction and conversion of dwellings, residential nursing homes, re-integration and specialist nursing homes • 30–60 % funding
Berlin	Investitionsbank Berlin IBB	• Assistance for dwellings to suit the needs of the elderly
Brandenburg	Investitionsbank des Landes ILB	• Assistance with barrier-free access • Lift installation programme directive
		• Modification of rented accommodation to suit the needs of the disabled
Bremen	Bremer Aufbaubank	• Increased loans for the additional costs of special constructional measures to suit the needs of the severely disabled
Hamburg	Wohnungsbau-kreditanstalt	Conversions of and extensions to dwellings (rented and owner-occupied) to suit the needs of wheelchair users and the elderly, with the following measures: • all constructional and technical measures to achieve a dwelling meeting the needs of the disabled • conversion measures to create a dwelling meeting the needs of wheelchair users (Model F) • retrofitting of lifts in tenanted multi-occupancy buildings (Model G)
Hesse	Landestreuhand-stelle Hessen LTH	• Assists with rented accommodation on the basis of the Housing Assistance Act (*Wohnraumförderungsgesetz*, WoFG) • Assists with measures for barrier-free building and the procurement of dwellings for older people with care services
		• Assists with the modernisation of rented accommodation when there is considerable need to adapt to a changed lifestyle
	Local authorities	• Subsidies for eliminating constructional obstacles for disabled persons in existing dwellings and the immediate environment
Mecklenburg-Western Pomerania	Landesförderinstitut Mecklenburg-Vorpommen	• Assists with existing rented accommodation and cooperative-owned housing with care services
		• Loans for adapting dwellings to suit the needs of disabled or elderly persons
Lower Saxony	Investitions- & Förder-bank Niedersachsen Nbank	• Assists with the procurement of rented accommodation for groups of elderly or severely disabled persons or those in need of care or assistance
North Rhine-Westphalia	NRW.Bank	• Assists with the new construction, purchase or retrofitting of rented accommodation on the basis of a disability (assistance programmes: "New rented accommodation", "Creating new rented accommodation in the building stock")
		• Modernisation programme on the basis of the NRW assistance directive
		• Communal assistance programmes for conversion and adaptation measures
		• Assists with lifts for publicly assisted housing for elderly or disabled persons
		• Residential homes programme (new, upgrades, extensions) for disabled persons, especially constructional adaptation and modernisation work in existing homes for the elderly and nursing homes, and the provision of new nursing home places in conjunction with the programme to create rented accommodation plus new, upgrading and extension work to living quarters in homes for disabled persons
Rhineland Palatinate	Rhineland Palatinate Ministry of Finance	• Assists with projects that consist of up to 60 % residential accommodation in order to promote communal lifestyle forms with young and old people plus the mixing of living and working environments
		• Federal state programmes for publicly assisted housing
Saarland	Saarländische Investitions-kreditbank	• Assists with the modernisation of rented accommodation in order to reduce constructional barriers
Saxony	Sächsische Aufbaubank SAB	• Housing assistance within the scope of owner-occupied schemes
		• Housing assistance within the scope of the multi-generation housing scheme
Saxony-Anhalt	Investitionsbank Sachsen-Anhalt	• Assists with the expenditure of clients who pursue non-profit-making objectives for new construction, conversions and extensions, the purchase of buildings and plots plus interior fitting-out for creating establishments for disabled persons
		• Refurbishment, modernisation and repair of the existing housing stock in barrier-free rented accommodation
Schleswig-Holstein	Investitionsbank Schleswig-Holstein	• Assistance with rented accommodation for persons on low incomes, especially families, single-parent families, the elderly and the disabled
		• Retrofitting of lifts in rented accommodation
		• Assists with the new construction and modernisation of rented accommodation and cooperative-owned housing to form apartments suitable for the needs of the elderly
Thuringia	Aufbaubank Thüringen	• Housing assistance schemes within the scope of the inner-city stabilisation programme
		• Assists with interior fitting-out to create environments barrier-free and suitable for the elderly within the scope of new construction, conversion and extension works

Regulatory principles
Standardisation

T5: Federal state building regulations [4]

Land	Legislation, statutory instruments	Content
Baden-Württemberg	Federal State Building Regulations (*Landesbauordnung*, LBO)	• cl. 3 General requirements (4) • cl. 29 Lift installations (2) • cl. 35 Dwellings • cl. 39 Barrier-free facilities
	Federal State Disabled Persons Equality Act (*Behindertengleichstellungsgesetz*, L-BGG)	• cl. 3 Freedom from barriers • cl. 7 Barrier-free design and construction in buildings and transport • cl. 12 Right to legal action
	Garages Act (*Garagenverordnung*, GaVO)	• cl. 4 Parking places and access lanes
	Retail Premises Act (*Verkaufsstättenverordnung*, VkVO)	• cl. 27 Fire protection regulation (1) • cl. 28 Parking places for disabled persons
	Places of Assembly Act (*Versammlungsstättenverordnung*, VStättVO)	• cl. 10 Seating, aisles and stepped aisles (7) • cl. 12 WC facilities (2) for wheelchair users • cl. 13 Parking places for disabled persons • cl. 42 Fire protection regulation, fire brigade plans (1) Measures for rescuing disabled persons, especially wheelchair users • cl. 44 Building documentation, seating and escape/rescue routes plan (3)
Bavaria	Bavarian Building Regulations (*Bayerische Bauordnung*, BayBO)	• art. 32 Stairs • art. 37 Lifts • art. 48 Barrier-free building
	Bavarian Disabled Persons Equality Act (*Bayerisches Behindertengleichstellungsgesetz*, BayBGG)	• art. 4 Freedom from barriers • art. 10 Barrier-free design and construction in buildings and transport • art. 16 Right to class action
	Garage Parking Act (*Garagenstellplatzverordnung*, GaStellV)	• cl. 4 Parking place
	Retail Premises Act (*Verkaufsstättenverordnung*, VkV)	• cl. 27 Fire protection regulation • cl. 28 Parking places for disabled persons
	Places of Assembly Act (*Versammlungsstättenverordnung*, VStättV)	• cl. 10 Seating, aisles and stepped aisles (7) Spaces for wheelchair users • cl. 12 WC facilities (2) for wheelchair users • cl. 13 Parking places for disabled persons • cl. 42 Fire protection regulation, fire brigade plans (1) Measures for rescuing disabled persons, especially wheelchair users • cl. 44 Building documentation, seating and escape/rescue routes plan (3)
Berlin	Berlin Building Regulations (*Bauordnung Berlin*, BauO Bln)	• cl. 2 Terminology (10, 12) • cl. 39 Lifts (4) • cl. 49 Dwellings (2) • cl. 50 Parking places • cl. 51 Barrier-free building • cl. 52 Special structures (1)
	Federal State Equality Act (*Landesgleichberechtigungsgesetz*, LGBG)	• cl. 1 Equality decree • cl. 2 Discrimination prohibition • cl. 9 Guarantee of mobility • cl. 15 Extraordinary right to legal action
	Restaurants Act (*Gaststättenverordnung*, GastV)	• Supplemented by circular SenWiArbFrau II E No. 4/2006, "Freedom from barriers in restaurants"
	Operation of Constructed Facilities Act (*Betriebs-Verordnung*, BetrVo)	• Part I Publicly accessible buildings used by disabled persons in wheelchairs • Part III Fire safety inspections and monitoring of facilities: cl. 5 • Part IV Building-related operating regulations: Section 1: Retail premises: cl. 10 Fire protection regulation Section 2: Places of accommodation: cl. 15 Unobstructed escape/rescue routes, fire protection regulation, persons responsible; cl. 16 Barrier-free rooms Section 3: Garages: cl. 21 Designated parking places for motor vehicles Section 4: Places of assembly: cl. 26 Places for visitors according to seating and escape/rescue routes plan; cl. 36 Fire protection regulation, fire brigade plans
Brandenburg	Brandenburg Building Regulations (*Brandenburgische Bauordnung*, BbgBO)	• cl. 34 Lifts (5) • cl. 41 Dwellings (5) • cl. 45 Barrier-free building
	Brandenburg Disabled Persons Equality Act (*Brandenburgisches Behindertengleichstellungsgesetz*, BbgBGG)	• cl. 4 Freedom from barriers • cl. 10 Right to class action
	Brandenburg Garages & Parking Places Act (*Brandenburgische Garagen- & Stellplatzverordnung*, BbgGStV)	• cl. 5 Parking places for motor vehicles for specific groups of persons
	Brandenburg Retail Premises Building Act (*Brandenburgische Verkaufsstätten-Bauverordnung*, BbgVBauV)	• cl. 27 Fire protection regulation • cl. 29 Parking places for disabled persons
	Brandenburg Places of Assembly Act (*Brandenburgische Versammlungsstättenverordnung*, BbgVStättV)	• cl. 10 Seating, aisles and stepped aisles (7) • cl. 12 WC facilities (2) • cl. 13 Parking places for disabled persons • cl. 42 Fire protection regulation, fire brigade plans (1) Measures for rescuing disabled persons, especially wheelchair users • cl. 44 Building documentation, seating and escape/rescue routes plan (3)
Bremen	Bremen Building Regulations (*Bremische Landesbauordnung*, BremLBO)	• cl. 38 Lifts (7, 8) • cl. 47 Dwellings • cl. 52 Special buildings and rooms for special uses • cl. 53 Buildings for specific groups of persons

Regulatory principles
Standardisation

Land	Legislation, statutory instruments	Content
Bremen (contd.)	Bremen Disabled Persons Equality Act (*Bremisches Gesetz zur Gleichstellung von Menschen mit Behinderung*, BremBGG)	• cl. 4 Freedom from barriers • cl. 8 Barrier-free design and construction in buildings and transport • cl. 12 Right to class action
	Bremen Garages & Parking Places Act (*Bremische Verordnung über Garagen & Stellplätze*, BremGaVo)	• cl. 4 Parking places and traffic areas (1)
Hamburg	Hamburg Building Regulations (*Hamburgische Bauordnung*, HBauO)	• cl. 37 Lifts • cl. 52 Barrier-free building
	Hamburg Disabled Persons Equality Act (*Hamburgisches Gesetz zur Gleichstellung behinderter Menschen*, HmbGGbM)	• cl. 4 Freedom from barriers • cl. 7 Barrier-free design and construction in buildings and transport • cl. 12 Right to class action
	Garages Act (*Garagenverordnung*, GaVO)	• cl. 10 Parking places and traffic areas (1)
	Retail Premises Act (Verkaufsstättenverordnung, VkVO)	• cl. 27 Fire protection regulation (1)
	Places of Assembly Act (Versammlungsstättenverordnung, VstättVO)	• cl. 10 Seating (7) • cl. 12 WC facilities (2) • cl. 42 Fire protection regulation, fire brigade plans (1) Measures for rescuing disabled persons, especially wheelchair users • cl. 44 Building documentation, seating and escape/rescue routes plan (3)
	Underground Railway Building Directive	• Tactile floor elements
Hesse	Hesse Building Regulations (*Hessische Bauordnung*, HBO)	• cl. 33 Lifts • cl. 46 Barrier-free building
	Hesse Disabled Persons Equality Act (*Hessisches Behindertengleichstellungsgesetz*, HessBGG)	• cl. 3 Freedom from barriers • cl. 10 Barrier-free design and construction in buildings and transport • cl. 17 Right to class action
	Garages Act (*Garagenverordnung*, GaVO)	• cl. 5 Parking places and access lanes
	Model Places of Accommodation Act (*Muster-Beherbergungsverordnung*, M-BeVO)	• The M-BeVO is applied by the building authority when assessing places of accommodation
	Model Places of Assembly Act (*Muster-Versammlungsstättenverordnung*, MVStättV)	• Handled as for the M-BeVO • cl. 10 Seating, aisles and stepped aisles (7) • cl. 12 WC facilities (2) • cl. 13 Parking places for disabled persons • cl. 42 Fire protection regulation, fire brigade plans (1) Measures for rescuing disabled persons, especially wheelchair users • cl. 44 Building documentation, seating and escape/rescue routes plan (3)
Mecklenburg-Western Pomerania	Mecklenburg-Western Pomerania Building Regulations (*Landesbauordnung Mecklenburg-Vorpommern*, LbauO M-V)	• cl. 39 Lifts (5) • cl. 50 Barrier-free building
	Federal State Disabled Persons Equality Act (*Landesbehindertengleichstellungs-Gesetz*, LBGG M-V)	• cl. 6 Freedom from barriers • cl. 8 Barrier-free design and construction in buildings and transport
	Garages Act (*Garagenverordnung*, GaVO)	• cl. 4 Parking places and access lanes (4)
	Retail Premises Act (*Verkaufsstättenverordnung*, VkVO)	• cl. 28 Parking places for disabled persons
	Places of Assembly Act (*Versammlungsstättenverordnung*, VstättVO)	• cl. 10 Seating (7) • cl. 12 WC facilities (2) for wheelchair users • cl. 13 Parking places for disabled persons • cl. 42 Fire protection regulation, fire brigade plans (1) Measures for rescuing disabled persons, especially wheelchair users • cl. 44 Building documentation, seating and escape/rescue routes plan (3)
Lower Saxony	Lower Saxony Building Regulations (*Niedersächsische Bauordnung*, NbauO)	• cl. 36 Lift installations (3) • cl. 44 Dwellings (3) • cl. 45 WC facilities and bathrooms (3) • cl. 48 Barrier-free accessibility and usability of certain buildings
	Lower Saxony Disabled Persons Equality Act (*Niedersächsisches Behindertengleichstellungsgesetz*, NBGG)	• cl. 7 Barrier-free design and construction in buildings and transport • cl. 13 Right to class action
	Garages Act (*Garagenverordnung*, GaVO)	• cl. 4 Parking places and traffic areas (1)
	Retail Premises Act (*Verkaufsstättenverordnung*, VkVO)	• cl. 28 Parking places for disabled persons
	Places of Assembly Act (*Versammlungsstättenverordnung*, VstättVO)	• cl. 10 Seating, aisles and stepped aisles (7) • cl. 12 WC facilities (2) • cl. 13 Parking places for disabled persons • cl. 42 Fire protection regulation, fire brigade plans (1) Measures for rescuing disabled persons, especially wheelchair users • cl. 44 Building documentation, seating and escape/rescue routes plan (5)
North Rhine-Westphalia	North Rhine-Westphalia Building Regulations (*Bauordnung für das Land NRW*, BauO NRW)	• cl. 39 Lifts (6, 7) • cl. 49 Dwellings (2, 5) • cl. 55 Freedom from barriers in publicly accessible buildings with commentaries on the legal prerequisites of BauO NRW cl. 55
	North Rhine-Westphalia Disabled Persons Equality Act (*Behindertengleichstellungsgesetz NRW*, BGG NRW)	• cl. 4 Freedom from barriers • cl. 6 Right to class action • cl. 7 Barrier-free design and construction in buildings and transport
	Garages Act (*Garagenverordnung*, GaVO)	• cl. 6 Parking places and traffic areas (1)
	Retail Premises Act (*Verkaufsstättenverordnung*, VkVO)	• cl. 26 Parking places for disabled persons
	Places of Assembly Act (*Versammlungsstättenverordnung*, VstättVO)	• cl. 10 Seating, aisles and stepped aisles (7) • cl. 12 WC facilities (2) • cl. 13 Parking places for disabled persons • cl. 42 Fire protection regulation, fire brigade plans (1) Measures for rescuing disabled persons, especially wheelchair users

Regulatory principles
Standardisation

(T5 contd: Federal state building regulations [4])

Land	Legislation, statutory instruments	Content
Rhineland Palatinate	Rhineland Palatinate Building Regulations (*Landesbauordnung Rheinland-Pfalz*, Pfalz LBauO)	• cl. 36 Lifts (5) • cl. 44 Dwellings (2) • cl. 51 Buildings for specific groups of persons
	Federal State Disabled Persons Equality Act (*Landesgesetz zur Herstellung gleichwertiger Lebensbedingungen für Menschen mit Behinderung*, LGGBehM)	• cl. 2 Terminology • cl. 9 Barrier-free design and construction in buildings and transport • cl. 10 Right to class action
	Garages Act (*Garagenverordnung*, GaVO)	• cl. 4 Garage parking places, access lanes (1)
	Retail Premises Act (*Verkaufsstättenverordnung*, VkVO)	• cl. 28 Parking places for disabled persons
Saarland	Saarland Building Regulations (*Landesbauordnung*, LBO)	• cl. 39 Lifts (5) • cl. 50 Barrier-free building
	Saarland Disabled Persons Equality Act (*Saarländisches Behindertengleichstellungsgesetz*, SBGG)	• cl. 2 Terminology • cl. 10 Barrier-free design and construction in buildings and transport • cl. 14 Right to class action
	Garages Act (*Garagenverordnung*, GaVO)	• cl. 4 Parking places and traffic areas (1)
	Retail Premises Act (*Verkaufsstättenverordnung*, VkVO)	• cl. 28 Parking places for disabled persons
	Places of Assembly Act (*Versammlungsstättenverordnung*, VstättVO)	• cl. 10 Seating, aisles and stepped aisles (7) • cl. 12 WC facilities (2) • cl. 13 Parking places for disabled persons • cl. 42 Fire protection regulation, fire brigade plans (1) Measures for rescuing disabled persons, especially wheelchair users • cl. 44 Building documentation, seating and escape/rescue routes plan (5)
Saxony	Saxony Building Regulations (*Sächsische Bauordnung*, SächsBO)	• cl. 39 Lifts (4, 5) • cl. 50 Barrier-free building
	Saxony Integration Act (*Sächsisches Integrationsgesetz*, SächsIntegrG)	• cl. 3 Freedom from barriers • cl. 9 Representation by an organisation
	Saxony Garages Act (*Sächsische Garagenverordnung*, SächsGarVO)	• cl. 4 Parking places and access lanes (1)
	Saxony Retail Premises Building Directive (*Sächsische Verkaufsstättenbaurichtlinie*, SächsVerkBauR)	• 3.4 Parking places for disabled persons
	Saxony Places of Assembly Act (*Sächsische Versammlungsstättenverordnung*, Sächs VstättVO)	• cl. 10 Seating, aisles and stepped aisles (7) • cl. 12 WC facilities (2) • cl. 13 Parking places for disabled persons • cl. 42 Fire protection regulation, fire brigade plans (1) Measures for rescuing disabled persons, especially wheelchair users • cl. 44 Building documentation, seating and escape/rescue routes plan (5)
Saxony-Anhalt	Saxony-Anhalt Building Regulations (*Bauordnung Sachsen-Anhalt*, BauLSA)	• cl. 38 Lifts (4, 5) • cl. 49 Barrier-free building
	Saxony-Anhalt Disabled Persons Equality Act (*Behindertengleichstellungsgesetz Land Sachsen-Anhalt*, BGStG LSA)	• cl. 2 Terminology • cl. 6 Planning, coordination and advice • cl. 17 Right to legal action
	Garages Act (*Garagenverordnung*, GaVO)	• cl. 5 Parking places and access lanes, barrier-free building
	Places of Assembly Act (*Versammlungsstättenverordnung*, VstättVO)	• cl. 10 Seating, aisles and stepped aisles (7) • cl. 12 WC facilities (2) • cl. 13 Parking places for disabled persons • cl. 42 Fire protection regulation, fire brigade plans (1) Measures for rescuing disabled persons, especially wheelchair users • cl. 44 Building documentation, seating and escape/rescue routes plan (5)
Schleswig-Holstein	Schleswig-Holstein Building Regulations (*Landesbauordnung*, LBO)	• cl. 41 Lifts (5) • cl. 52 Dwellings (2, 5) • cl. 59 Barrier-free building
	Federal State Disabled Persons Equality Act (*Landesbehindertengleichstellungsgesetz*, LBGG)	• cl. 2 Terminology • cl. 3 Right to legal action • cl. 11 Barrier-free design and construction in buildings and transport
	Garages Act (*Garagenverordnung*, GaVO)	• cl. 5 Parking places and access lanes (1)
	Places of Assembly Act (*Versammlungsstättenverordnung*, VstättVO)	• cl. 10 Seating, aisles and stepped aisles (7) • cl. 12 WC facilities (2) • cl. 13 Parking places for disabled persons • cl. 42 Fire protection regulation, fire brigade plans (1) Measures for rescuing disabled persons, especially wheelchair users • cl. 44 Building documentation, seating and escape/rescue routes plan (5)
Thuringia	Thuringia Building Regulations (*Thüringische Bauordnung*, ThürBO)	• cl. 37 Lifts (4, 5) • cl. 53 Barrier-free building
	Thuringia Disabled Persons Equality & Integration Improvement Act (*Thüringer Gesetz zur Gleichstellung & Verbesserung der Integration von Menschen mit Behinderung*, ThürGIG)	• cl. 5 Freedom from barriers • cl. 10 Barrier-free design and construction in buildings and transport • cl. 20 Litigation protection by organisations (right to legal action)
	Thuringia Garages Act (*Thüringer Garagenverordnung*, ThürGarVO)	• cl. 4 Parking places and access lanes (1)
	Thuringia Retail Premises Act (*Thüringer Verkaufsstättenverordnung*, ThürVStVO)	• cl. 27 Fire protection regulation (1) • cl. 28 Parking places for disabled persons

In contrast to the standards, the building regulations of the federal states are legally binding statutory instruments. The building regulations of the federal states stipulate which buildings and establishments, or parts thereof, must be erected without barriers, those which must be repaired or under what circumstances deviations are possible. In the majority of Germany's federal states the technical realisation of freedom from barriers is either specified directly in the building regulations of the particular federal state or regulated in the technical construction regulations of the particular federal state which are adopted in building legislation.

Developments in society

Fortunately, medicine has made astonishing progress over the past century. The benefits include:
- Better hygiene
- Declining mortality rates among babies and infants
- Successes in controlling epidemics
- Favourable living conditions due to changing social orders and safeguards

Thanks to these improvements, these days more and more people in Europe are living longer while quite likely still remaining relatively healthy and independent.

However, as the number of older citizens rises constantly, so the number of births drops steadily, which means we are witnessing a dramatic shift in population pyramid structures. This "demographic change" not only affects the structure of society and its stability (e.g. the principle of the "inter-generational contract"); it also has a considerable knock-on effect for the investment, planning and construction sectors.

This shift in the age structure also changes the nature of the needs of building users. Although many older people are living longer independently and remaining in relatively good health, they must still learn to cope with physiognomic changes. As we get older, our mobility and perception capabilities are no longer necessarily available to the same extent as they were 10 or 20 years previously. Of course, this in no way means that old age must be equated with disability. Nevertheless, constructional conditions corresponding to barrier-free or universal design would represent a great help to this group of citizens in particular. For the 80+ age group, daily activities are three times more likely to be difficult to impossible compared to those aged 65–79 (Figs. 14 and 15). So this demographic development has an effect on the size and accessibility of dwellings on the whole, as well as on a whole series of constructional details, e.g. barrier-free access, door widths, sanitary facilities, access to balcony or patio.
If we widen our view to take in not only the growing population of older people but also those with disabilities, and embrace the idea of designing an environment without barriers for anybody, then it is worthwhile studying the figures of demographic developments.

Worldwide demographic developments

Whereas the average growth in the world's population has decreased over the past century, the situation at national level is very varied. Only a few countries in the world, e.g. Germany, Italy, Poland, are recording a declining population.

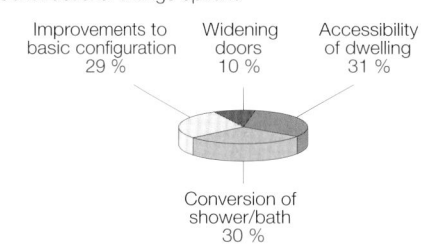

Constructional change options
Improvements to basic configuration 29 %
Widening doors 10 %
Accessibility of dwelling 31 %
Conversion of shower/bath 30 %

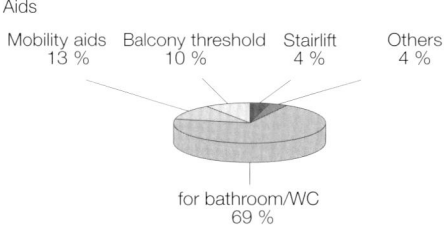

Aids
Mobility aids 13 %
Balcony threshold 10 %
Stairlift 4 %
Others 4 %
for bathroom/WC 69 %

Fittings
Living/sleeping areas 23 %
Kitchen 15 %
Others 20 %
Bathroom usage 42 %

The problems faced by older people at home

All figures in %	Aged 65–79	Aged 80 and over
Climbing stairs	13.5	33.5
Taking a bath	8.9	30.2
Preparing meals	5.9	24.8
Showering/washing	5.7	21.9
Heating the dwelling	5.5	19.8
Walking around	3.6	14.3
Going to bed/getting up	3.5	13.2
Sitting down/getting up	3.6	12.3
Using the toilet	2.4	10.8

14 The problems faced by older people in performing daily activities at home; Federal Ministry for Families, Senior Citizens, Women & Juveniles (BMFSFJ)
15 Options for constructional changes, NRW model programme 1994–95

Developments in society
Worldwide demographic developments

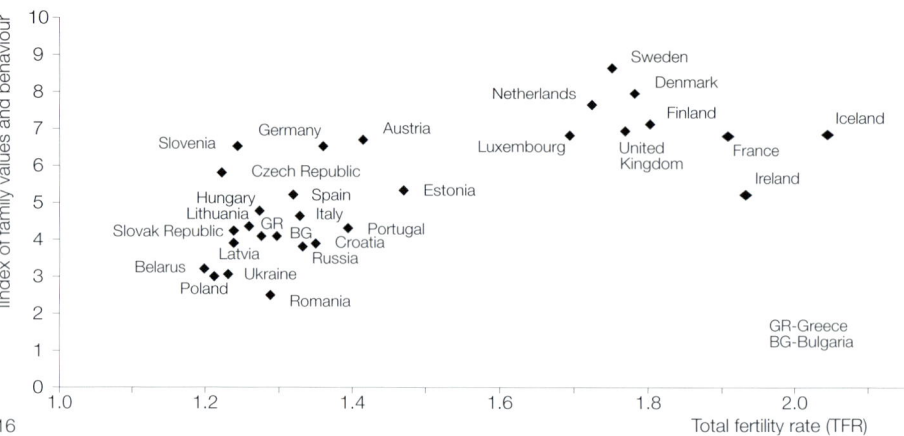

16

The figures that Mary Mederios Kent and Carl Haub of the Rostock Centre for Research into Demographic Change released in December 2005 [1] show that falling birth rates is not a global phenomenon. The only continent that will probably see a decline in its population is Europe (Fig. 17).

The expression "demographic divide" describes the gulf between birth and death rates in individual countries. On the one side we have the majority of rich countries, whose birth rates are decreasing and in which, at the same time, the average life expectancy of 75 and higher is being increased by the relative proportion of older people. On the other side we have the relatively poor "least developed countries" (LDC countries) with their relatively high birth rates but low life expectancies.

The demographic prospect in Europe
At the end of 2008 the Max Planck Institute for Demographic Research published figures that confirm the above predictions. [2] Three-quarters of the population of Europe lives in countries in which birth rates of less than 1.5 children per couple have been verified. The study shows that not one single country reaches the "replacement level" of, on average, 2.1 births per woman, i.e. the level at which a couple has only enough children to replace themselves (Fig. 16).

There are many reasons for this development, including the fact that couples start having families later because they spend more time in education and training, and the number of women in full-time employment has increased.

Demographers assume that these developments will not remain a transient phenomenon restricted to certain geographical regions. Central, eastern and southern Europe are particularly affected, while the birth rates in the countries of northern and western Europe are slightly higher. The ageing of the population on the whole will also continue.

In order to slow down this process at least, the researchers have the following recommendations:
• Material inducements through family policies
• Better childcare services
• More flexible working hours
• Promoting equality between men and women
• Changes to migration policies

Demographic change in Germany
The birth rate will drop again in the future, and the consequence of that will be that the number of potential mothers will decline as well. Currently, the number of girls being born every year is already lower than that of their mothers.
Despite the increased life expectancy, death rates will rise when those born in the years with high birth rates reach old age. This number will increasingly exceed the number of births. The net immigration rate will no longer compensate for the rapidly increasing births deficit. As a result, the

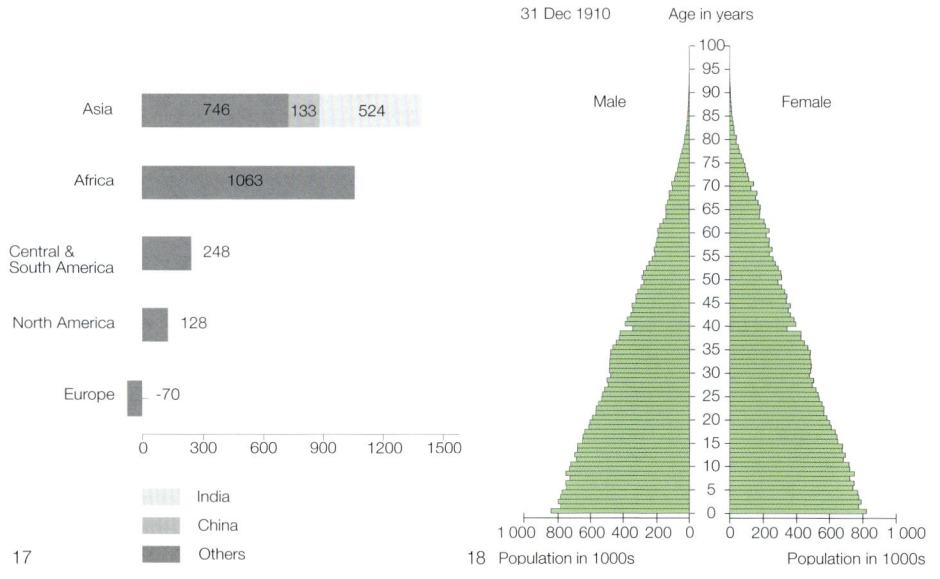

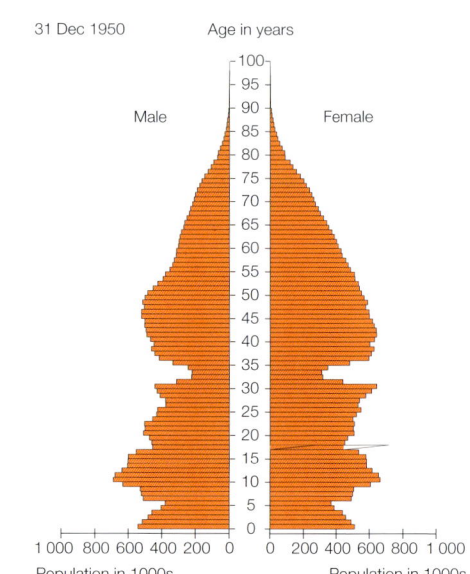

17
18

30

Developments in society
Demographic change in Germany

population of Germany, which has been falling since 2003, will continue to decrease. If the current demographic trend continues, the population of almost 82.5 million in 2005 will have diminished to at best 74 million, but possibly just under 69 million, by 2050 (Fig. 18). [3]

The relations between young and old will alter considerably. At the end of 2005 some 20 % of the population were younger than 20, whereas the 65+ age group accounted for 19 %. Persons in the "working age" group (20 to 64) accounted for the remaining 61 %. At the same time, the group of the population aged 80+ has been and will continue increasing relentlessly: from just under 4 million in 2005 to 10 million in 2050. By that date more than 40 % of the 65+ age group will be at least 80 years old. By contrast, in the year 2050 only about half the population will fall within the working age bracket. More than 30 % will be 65 or older and approx. 15 % will be < 20. Currently, for every 100 persons of working age there are 33 below the age of 20. This ratio will only decrease marginally and by 2050 will have fallen to 29. In the future the group of pensioners will grow with respect to the population of working age. In 2005 there were 32 pensioners for every 100 persons of working age. By 2030 this ratio will have increased to 50 or 52, and in 2050 it will be 60 or 64 (Fig. 19). Even if the retirement age is raised, the ratio of the 67+ age group in 2050 will still be much higher than that of the 65+ age group today.
The overall ratio of those persons who have not yet reached working age, or are no longer of working age, to those of working age will be dominated by the increase in older people. In 2005 there were 65 people ≤ 20 and ≥ 65 for every 100 persons between 20 and 65; by 2030 this figure will have risen to > 80 and in 2050 it will be 89 or 94. [3]

Developments in private households
The tendency in Germany is for shrinking

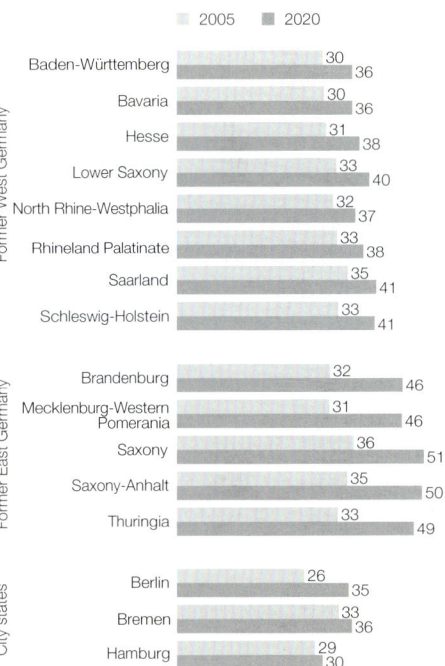

19 Ratio of pensioners to those of working age for 2005 & 2020
No. of persons aged 65 and over
for every 100 persons between the ages of 20 and 64

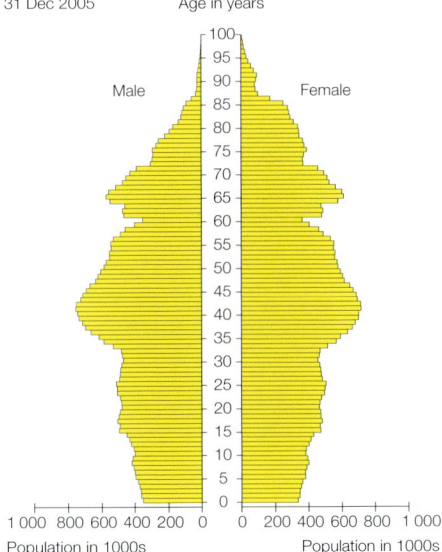

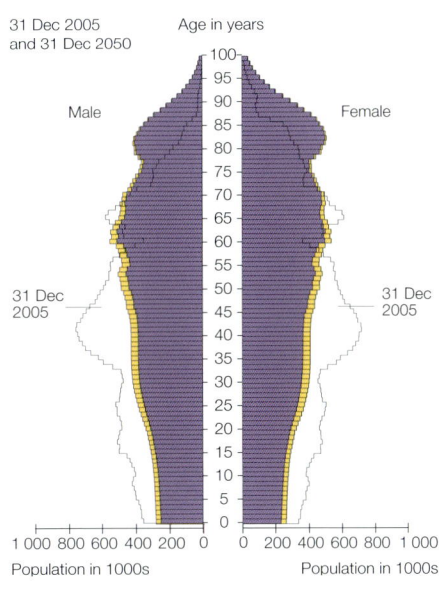

16 Index of family values and behaviour plotted against the total fertility rate in the respective country (2004); source: project report, chap. 6, Frejka, 2008
17 Predicted population growth or decline according to continents and selected countries for the period 2005–2050; Carl Haub, 2005 World Population Data Sheet, zdwa
18 The age structure of the population of Germany up until 2050; 11th Coordinated Population Prediction, Federal Statistical Office, 2006
  ■ lower limit of "mean" population
  ■ upper limit of "mean" population
19 Ratio of pensioners to those of working age for the years 2005 and 2020; results of 11th Coordinated Population Prediction, Statistical Offices of the Federation and the Länder, 2006

Developments in society
Consequences for developments in urban planning and the built environment

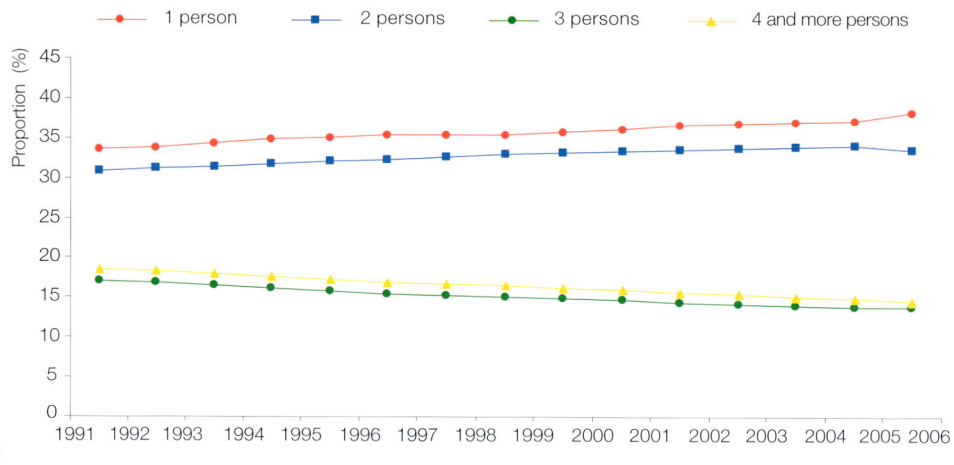

20

households. This trend has been observed since statistics started to be collected in the late 1950s.
One feature in this development over the last 30 years is that the number of single-person households now exceeds all other household sizes. This share is growing steadily and it is not always younger people who are living alone; there are more older people, too. The number of two-person households is also on the increase. But households with three or more persons are gradually declining (Fig. 20). [4]
A continuation of the trend towards smaller households is expected for the future. The following factors indicate that this tendency will lead to more single- and two-person households for older people as well:
• Declining death rates
• The ever higher life expectancy of women
• The faster increase in the life expectancy of men compared to that of women

There were 39.4 million households in Germany in 2005 – an increase of 12 % since 1991. However, the trend towards smaller households leads to their numbers developing differently to that of the population.
Between 1991 and 2005 the number of households rose four times faster than the number of people living in them. Up until 2020 the population living in private households will probably decrease by 3 %, whereas the number of households will increase by 3 %. Therefore, if the trend continues, the forecast is 40.5 million households in Germany in 2020.

Single-person households will climb 9 % from the current figure of 15 million to reach 16.5 million; the increase in the number of two-person households will be even greater: from 13.2 million to 14.7 million, a rise of 11 %. In contrast, the number of larger households is expected to decrease: from 5.4 to 4.7 million (13 % decline) for three-person households and from 5.7 to 4.6 million (19 % decline) for households with four or more people.

There are 8.6 million persons living in Germany who are officially registered as disabled, and some 8.4 million of them in private households. Some 28 % of this group lives alone, just over half live in partnerships and 22 % form communities with three or more persons.

The structure of the dwellings is changing dramatically as well as the sizes of the households. The living space per person in Germany has almost doubled in just a few decades: from 22 m^2/person in 1966 to 42 m^2/person in 2007. At the same time, both the number of occupants and their comfort expectations have changed.

All these changes predicted for the near future represent major challenges for local communities and the housing market in particular. For example, the predicted change in the make-up of the population will require a corresponding infrastructure tailored to its needs. This infrastructure includes all areas of direct general services (foodstuffs, daily commodities, etc.) as well as, for example, local healthcare services, local public transport, access to media, etc. Although politics proclaims equality between living spaces, it is almost impossible to imagine achieving this in practice. The regional differences and opportunities are too pronounced.

Consequences for developments in urban planning and the built environment
All these figures and definitions are much more than just ends in themselves, are

20 Households according to size in Germany; Statistical Offices of the Federation and the Länder, 2007
21 Age-related frequency of dementia; source: Merz Pharmaceuticals, 2009

Developments in society
Consequences for developments in urban planning and the built environment

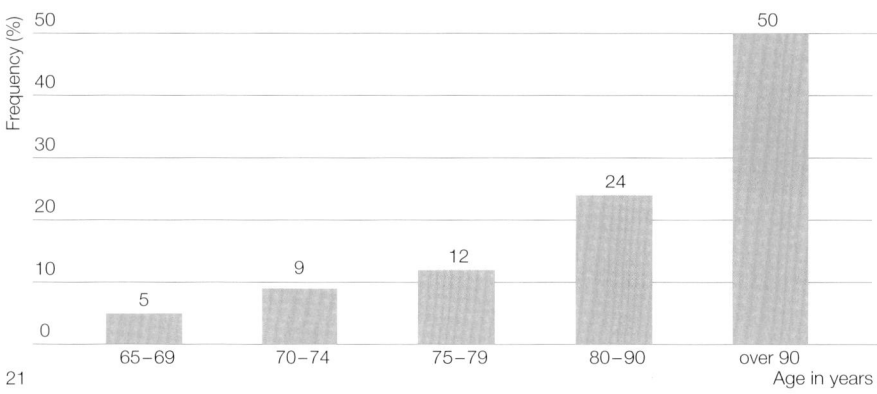

21

not merely empirical confirmation of suspected developments in society. Rather, they enable abstractions that make orientation and allocation even more comprehensible. They can support urban, regional and development planning, in conjunction with the specific local circumstances, to the same extent as political decisions regarding families, society or migration. Besides fundamental issues concerning energy-efficiency upgrades to and barrier-free refurbishment of the building stock, we must find local answers to different developments. The challenges facing growing cities such as Hamburg, Frankfurt or Munich cannot be compared with the needs of shrinking regions.

At the moment there is a shortage of housing tailored to the needs of the elderly. The German Association of Housing & Property Companies (Bundesverband deutscher Wohnungs- & Immobilienunternehmen) published figures in 2004 that show that of the 39 million dwellings in Germany, just 350,000 meet the needs of older people! And the German Association of Independent Property & Housing Companies (Bundesverband Freier Immobilien- & Wohnungsunternehmen) is predicting that by 2020 we will need an additional 800,000 new or appropriately modernised dwellings. [5] The actual need for care facilities and their specific orientation is currently the subject of heated debate, primarily because those persons who fall within care categories 0 and 1 will in future no longer be looked after in residential homes. Irrespective of these structural issues, the following figures, published by the Federal Ministry of Labour & Social Affairs in July 2009, should be taken into account:
In 2005 there were 1.2 million people in Germany suffering from dementia. The Ministry expects this figure to increase to more than 2 million by 2050. Dementia is the most common and the most serious psychiatric disorder affecting older people. Those suffering from Alzheimer's disease experience loss of memory and speech. They become increasingly unable to plan and act and therefore rely on help (Fig. 21).

The objectives of building without barriers (accessibility, usability) will also have to become the generally applicable quality standards, corresponding to the needs of older people especially, in areas other than care facilities.
It has been proved many times in practice that early integration of the definitions in DIN 18024-2 "Publicly accessible buildings and workplaces, design principles" and DIN 18025-2 "Accessible dwellings; design principles" does not result in additional costs. [6]

Regrettably, the definitions for the abolition of barriers have also resulted in the tendency to create constraints such that in certain areas or for certain types of usage those affected are forced to behave in ways defined by others. [7]
If it was not already clear, the ambivalence of the efforts to abolish barriers becomes obvious here. On the one hand, individual functionality, personal room for action and movement is supported, on the other, differences, dependencies and the social pressure to conform become evident. [8]

It is not just social science studies that indicate that barriers do not simply exist, but are constantly practised and qualified. There are therefore no spaces that can be used and relied upon by everybody to the same extent.
But it is exactly here that we find the inexhaustible chances for differentiation and hence for orientation in the planning and designing of our built environment and planners' and designers' responses to each specific location.

[1] Mederios Kent/Haub, 2005, p. 4ff.
[2] Frejka et al., 2008
[3] Statistical Offices of the Federation and the Länder, 2006
[4] Statistical Offices of the Federation and the Länder, 2007
[5] Marx, 2009, p. 28ff.
[6] ibid., p. 32ff.
[7] Bösl, 2009
[8] ibid.

33

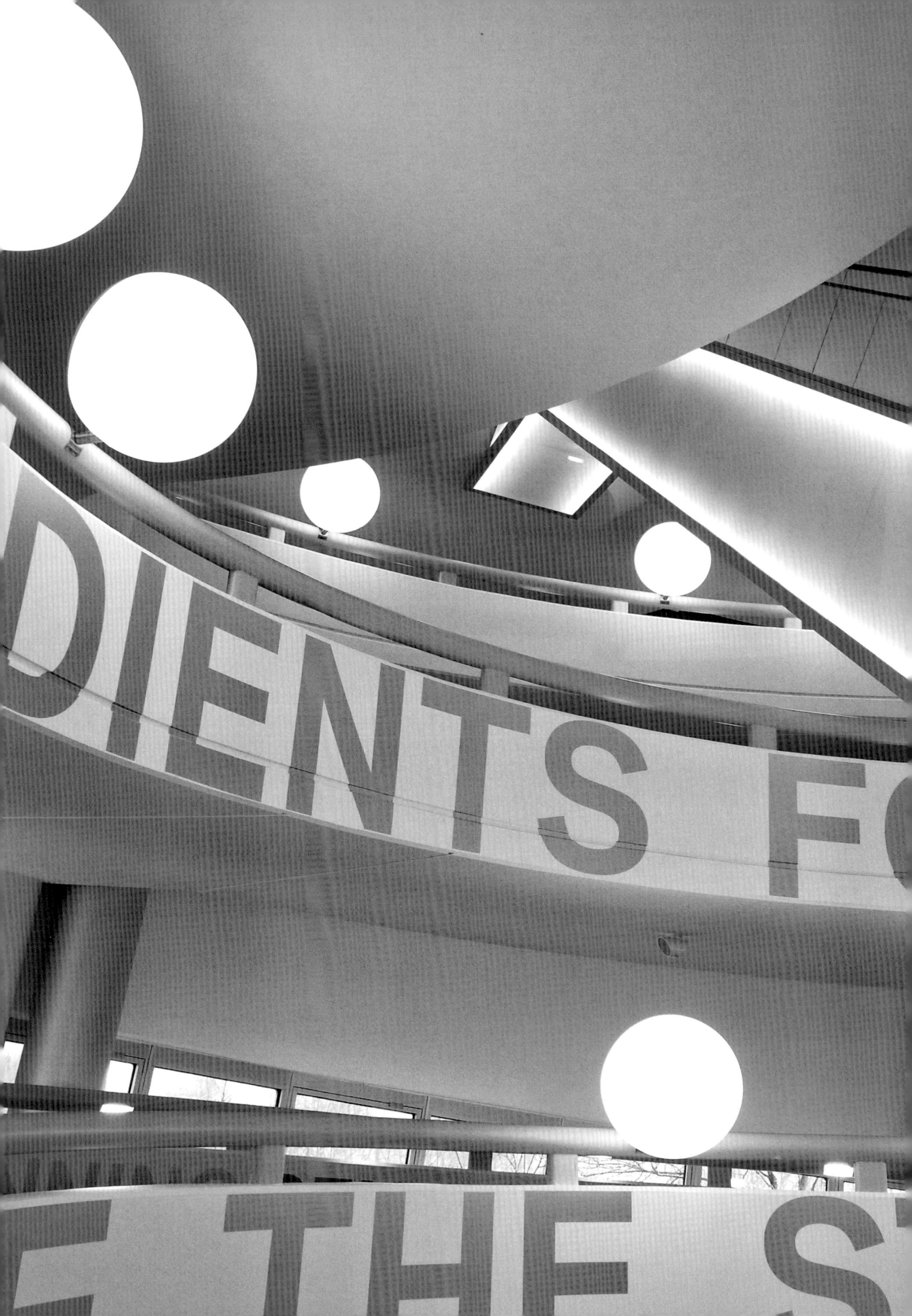

Design

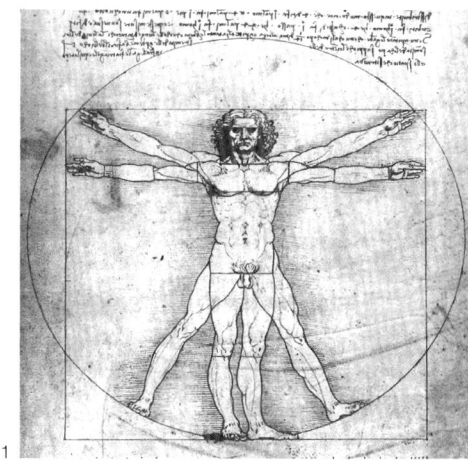

Ever since human beings have been able to produce tools as well as just use them, they have been designing artefacts based on dimensions. The Sophists of ancient Greece regarded the human being as the "measure of all things". This view has at least two meanings: it sees the human being as value and measure, and it formulates how the man-made environment depends on the dimensions of a human being. This depiction of the relative proportions and the way the human body is related to space, first described by Vitruvius around 25 BC, was famously illustrated by Leonardo da Vinci (Fig. 1). It is easy to understand how the first dimensional units were based on the human body (cubit, foot, hand's-breadth, "as tall as a man"). Despite the individuality of the human body, it is therefore possible – once we get over our initial shock – to specify this anthropometric data, the dimensions of the human body, in a standard, DIN 33402. Such a standard has been important since industry has been manufacturing everyday articles on a larger scale, e.g. furniture (desks, worktops, beds, chairs, washbasins, etc.), means of transport (cars, bicycles, buses, etc.) and built-in elements (vanity units, baths, taps, WCs, etc.) [1].

Irrespective of changing capabilities, skills and the different size relationships from childhood to old age, this data can alter owing to changes in our posture, the way we move or our strength. Dependencies or limitations can lead to us no longer being able to cope with our environment independently without certain aids (e.g. walking frame, wheelchair). When we take such aids into account, the areas we need for moving or passing, and our operating options and visibility ranges are affected (Fig. 2, p. 36; Figs. 4 and 5, p. 37).

Limitations and the ensuing demands on the designer

The aim of building without barriers is to make the built environment accessible and usable for people with impaired motoric capabilities or sensory organs. It would therefore seem reasonable to look at each of these aspects in more detail. Mobility and physical condition are the traditional areas that have been considered in architecture. Only recently have designers started to deal more intensively with design requirements resulting from limitations due to sensorial, i.e. seeing and hearing, and cognitive impairments.

As already explained in the first chapter, "On the history of barrier-free design and construction" (p. 9), disabilities can be categorised as follows:
- Motoric impairments (restricted movement, strength, sense of balance, dexterity and coordination abilities)
- Sensorial impairments (visual, hearing, lack of sense of smell or taste)
- Cognitive impairments (speech, learning or mental functions)
- Mental impairments

Motoric impairments
Motoric impairments (restrictions on bodily movements) are more readily perceived by the public than sensorial ones. Motoric problems lead to and are generally associated with mobility problems. However, the causes of mobility problems may also be found in age, accidents, illnesses or sensorial impairments.

1 Canon of Proportions, Vitruvian Man, Leonardo da Vinci, 1492

35

Design
Limitations and the ensuing demands on the designer

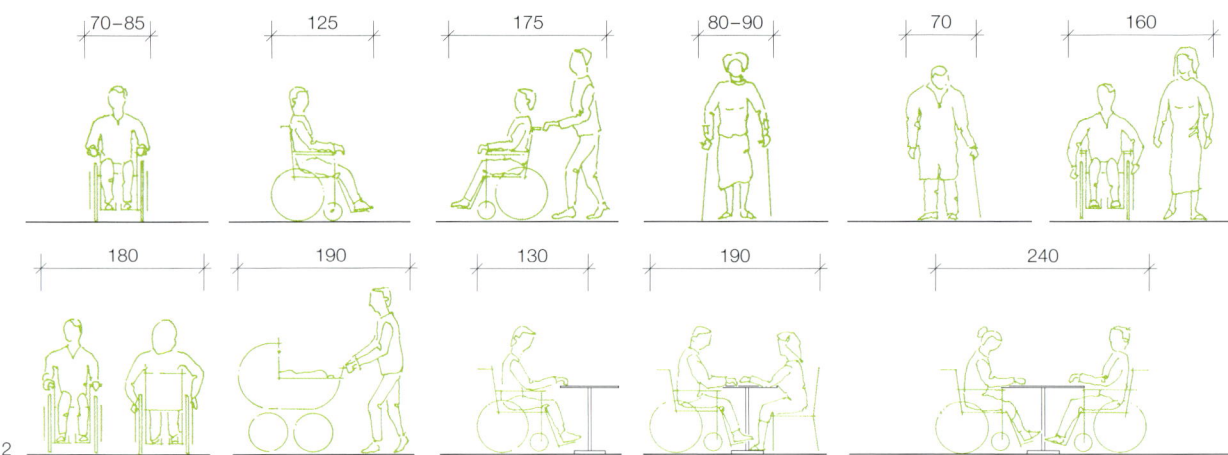

These functional deficiencies concerning our muscular-skeletal and locomotor systems may be caused by damage to the following:
- Brain
- Spinal cord
- Muscles or skeleton
- Limbs or lack thereof
- Functioning of internal organs

Such problems can lead to the following:
- Movement coordination disorders (spasticity, athetosis or ataxias)
- Partial or total paralysis
- Vegetative function or nerve deficiencies
- Muscular dystrophy or myasthenia, disruption of the muscle metabolism
- Impairments to holding, posture and walking functions
- Vertigo
- Any combination of these impairments

A barrier-free environment can try to compensate for motoric deficiencies and hence ease sequences of movements. Adequately sized movement areas and passing places, means of access of adequate width and without steps or thresholds plus even, non-slip floor coverings, handrails and short distances are the important factors to be considered if we are to ensure a certain degree of comfort and convenience for wheelchair users and those with mobility impairments. [2]

The shaping and individual form of handles, handrails, etc. are just as important when designing to avoid barriers. Vertigo or paralysis lead to mobility impairments. Walking frames, wheeled walkers or wheelchairs help restore static and dynamic mobility. Architecture without barriers, with unhindered internal and external circulation, takes account of the special needs of people with limited mobility by incorporating lifts or gentle ramps, for example.

Sensorial impairments
Initially, the human body itself tries to compensate for a deficit in sensory perception. Nature helps itself by intensifying the development of a "substitute sense". The "principle of two senses", a term coined only fairly recently, is a way of describing how with aids we can try to compensate for defective sensory perception or obtain other information from unaffected senses (see "Regulatory principles", p. 17).

We distinguish between the following priority levels with respect to the importance of information:
- Level 1: alarm and warning signals in the case of threats to life and limb, e.g. fire alarm
- Level 2: information that is important for making decisions and for which there is no option of further inquiry, e.g. signs in urban spaces
- Level 3: information with the option of further inquiry or supplementary data, e.g. the labelling on an exhibit at an exhibition

When it comes to vision, the functional impairments that affect human beings are very varied: they range from limited vision itself, i.e. short- or far-sightedness, to disruptions to the field of vision, e.g. "tunnel vision", deficiencies in sensing light or colour and focusing problems caused by other diseases. Impairments to or the loss of a sense should be regarded as totally different forms of disability. A blind person must receive different types of information to a person with some residual visual capacity. This also applies to people with hearing impairments: total deafness requires different measures to limited hearing capacity.

Suitable acoustic information systems help persons with impaired hearing to find their way around, but especially, to recognise dangers in good time. Acoustic environmental influences often have a negative effect on the communication options of this group of people. Disturbing or loud noises and poor room acoustics reduce their ability to perceive what is going on and hence affects the quality of their lives.

Contrasting structures and materials, e.g. in floor coverings, are an aid to tactile and haptic capabilities. The greater the differences, the easier it is to register these with a white cane or the feet.

Hearing
Persons with impaired hearing essentially use their residual ability for hearing. Amplifying hearing aids can provide assistance here: acoustic signals must therefore reach the hearing aid in the first place. Audio induction loops – wires concealed in the floor, ceiling or walls – have been fitted in many public areas. A special amplifier generates a magnetic field within the loop with the desired signals, i.e. music or speech. Through induction, this field generates a voltage in a second loop, located in the hearing aid, which in turn converts the signals back to their original form. In the open air, however, this principle is virtually unusable for technical reasons. Outside, it is better to work with a lower information input, according to the principle of two senses, i.e. by displaying texts on large screens – a solution that also helps those without any sense of hearing at all. This principle can be applied with dynamic information, too. The best solution is when the information is provided visually as well, by means of electronic displays. This is more effective and easier to understand than barely intelligible loudspeaker announcements, e.g. in railway stations or public buildings. The objective for all planners must be to provide suitable information in order to make mobility easier for all those people with impaired senses. For example, the

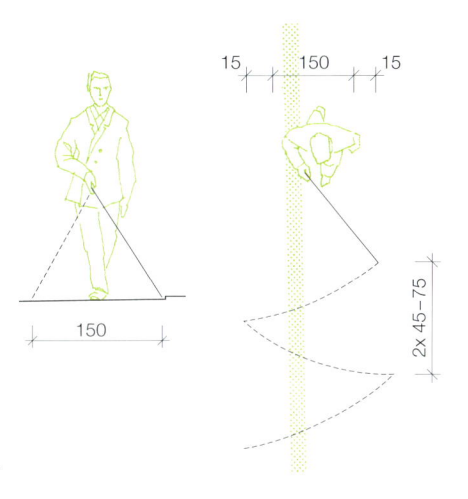

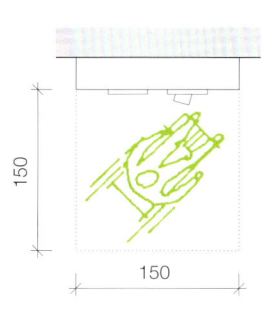

Design
Limitations and the ensuing demands on the designer

visually impaired must be provided with acoustic information in lifts, e.g. by announcing each floor.
In the event of dangers, alarms and warning signals must reach all people according to the principle of two senses, and must also enable anybody to warn others. If, for example, a deaf person discovers a fire, it will probably not be possible for that person to use a telephone or intercom to warn others or call the fire brigade. Sending a fax to the emergency services is currently not practicable because fax machines are not available to everybody. Experts and associations are therefore calling for this deficit to be rectified so that it is possible to warn others nationally and internationally by fax.
In lifts, for example, the intercom must be supplemented by an illuminated sign. Somebody stuck in a lift who triggers the alarm can therefore see that the alarm has been received and hence feel reassured. Many application options for priority level 2 are to be found in homes. Household appliances should ideally indicate their current status visually and acoustically. Televisions can work with subtitles for the hard-of-hearing and at the same time display a sign language interpreter for deaf viewers.
It is absolutely essential to improve the conditions for persons with sensorial impairments in places for public events (e.g. presentations, church services, lectures, etc.) (see "Typology", pp. 93–95). Modern halls and other places of assembly should include appropriate systems for transmitting the acoustic content to participants with hearing impairments. We distinguish between three transmission technologies depending on type of use and type of room:
- Inductive systems (audio induction loops)
- FM systems
- Infrared systems

In order that all persons can choose any seat, all audience zones in places of assembly should be equipped with audio systems and this fact appropriately indicated.
When using audio induction loops, the hearing aids pick up the signals (via the telecoil). Otherwise, additional receivers and neck loops or plug-in audio cables will be required.

Good visual equipment is also desirable. This could include a system that projects the speaker's mouth onto a large screen on the wall for lip-reading purposes. Overhead projectors, beamers and screens for digital presentations are usually standard these days, but the use of a spotlight on a sign language interpreter is a useful addition. Supplementary constructional measures are necessary in many instances. Those measures include adequate insulation against noise from outside and adjoining rooms. In rooms for the public, improved acoustics are necessary as well as sound insulation. The acoustic measures can include attenuating the noise caused by the audience itself. This is achieved, for example, by installing carpets or using wooden instead of steel furniture, as well as specific acoustic building measures. Short reverberation times are essential, echoes must be ruled out. Acoustic problems in particular are very complex and specialist consultants will probably have to be called in.

Seeing
Persons with impaired vision are particularly at risk in the streets, at transport stops and stations, and especially at the transitions between these elements. Added to that are situations with difficult lighting conditions. Paths, streets and open urban spaces must therefore be designed without obstacles and include appropriate areas for movement (see pp. 42–49 and Tab. T5, pp. 70–73). This means, for example, that no disturbing objects may intrude into paths. Changing levels and glass doors also represent potential hazards for the visually impaired. Contrasts can provide valuable help in such situations because the persons affected can therefore cope better and reach services and emergency facilities quickly without injuring themselves. When using visual guidance systems, the aim should be to ensure these are simple and readily legible when viewed close-up. Therefore, in surroundings with potential risks there should be no distracting advertising, whether in the form of posters or projected images. Shiny surfaces should also be avoided. Reflections can often impair a person's three-dimensional perception of a space and make orientation more difficult. Contrasts used sensibly with well-chosen colour combinations are fundamental to unambiguous perception – in conjunction with appropriate lighting. The sizes of objects or information are also relevant.

2 Dimensions for people with differing requirements. The figures given do not correspond to the movement and passing areas specified in the standard.
3 More and more older people are able to retain a certain independence by using mobility aids.
4 Movement areas for a visually impaired person who swings a white cane as an aid to orientation.
5 Movement area for a wheelchair user.

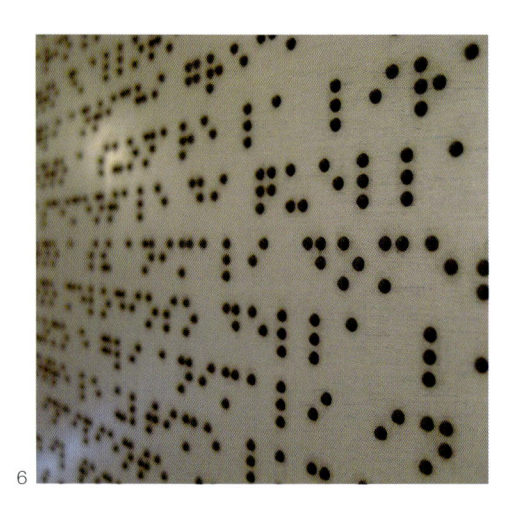

Design
Limitations and the ensuing demands on the designer

Contrast is defined as the difference in the luminance between neighbouring zones. It is a measure of the "perceived brightness". The human eye perceives differences in luminance as differences in brightness. This depends not only on the illuminance and the angle of incidence of the light, but also on the reflectance of the material or surface. Lighting that is optimised for restricted vision must provide an average luminance of 100 cd/m^2 as well as an even level of illumination. Optimum values lie between 250 and 300 cd/m^2. The difference in luminance between the brightest and darkest areas of a room should not be greater than 10:1, the difference in the immediate working environment, e.g. on a desk, should not exceed 3:1. To check the contrast, it is necessary to imagine the perceived image of the object as seen, not in colour, but exclusively in grey scale values, which is a much easier way of recognising differences in brightness. Values between 0 and 1 are used to define luminance contrasts. An extremely bright (white) object in front of a very dark (black) background will very probably score a value close to 1. Emergency call buttons and important warning signs, for example, must exhibit a contrast of at least 0.7. If the object is less strongly illuminated than its surroundings, we speak of a negative contrast. If it stands out from its background visually because of better luminance, the contrast is referred to as positive.

Proper illumination is essential for places that involve potential risks for disabled persons, e.g. stairs. With an inadequate level of illumination, a danger is either recognised too late or, at worst, not at all. Excessively bright sources of light can lead to distracting glare or strong shadows. Stairs placed in the middle of an area must also be indicated in some way in order to prevent persons from banging their heads on the underside of a flight (see Fig. 33, p. 50). When it comes to colour combinations, designers must consider that persons with disrupted colour vision cannot recognise certain colours. Owing to the relative frequency of red-green colour blindness, this combination is not to be recommended.

The materials of objects and their surroundings plus their surface textures also influence how we perceive contrast. The use of a conspicuous colour alone does not necessarily convey an important signal; nevertheless, certain colours have become traditional for certain situations:
- Red for prohibitions
- Yellow-black for warnings
- Green for OK

Recommendations are available for the sizes of lettering on information on signs or illuminated displays. In normal cases we can assume that information is at eye level, but where a sign is at a higher level, the lettering must be correspondingly larger. Legibility is also affected by other factors: the horizontal distance of the observer from the object, the angle of the sign on the wall or column and the type of sign. Sans serif typefaces are best, semibold or bold, using conventional upper- and lower-case letters.

The location itself also plays an important role. Signs at railway stations must be installed at a height that enables them to be readily visible from afar. More specific, additional information such as opening times or timetables are easier to read close-up when placed at a lower level. The space in front should therefore be kept clear of obstacles.

Visual information in all public areas, e.g. buildings to which the public has access, streets and public transport, must satisfy certain criteria. Those include good perceptibility and legibility, including from a distance, plus easy, unambiguous intelligibility. Signs should be installed in positions that suit their function. Information with a similar character should be grouped together. Where signs are intended to guide traffic flows, the information should be repeated at regular intervals.

Any obstacles, e.g. steps, that protrude into designated circulation zones, e.g. paths, must be marked in a contrasting way. The nosings of steps must be highlighted. Moving walkways, too, which are mostly found in airports, must be provided with markings so that users are warned even when their view forwards is restricted (see Tab. T5, p. 71, "Stairs").

6 Exhibition panel with tactile information on the exhibits in Braille, Pinacotheca, Vicenza (I)
7a, b Warning markings on glass doors; any form is possible, e.g. a company logo. Markings are positioned at knee and eye level. Bar-type markings (height approx. 6–8 cm) should extend across the full width of the door; height above FFL: 50 and 150 cm, ±10 cm
8a, b Door designs employing good contrasts
9 Examples of contrasting combinations of colours (a). A grey scale with light/dark contrasts is recommended for checking the contrast (b).

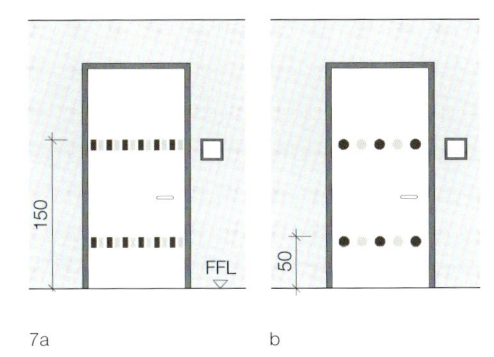

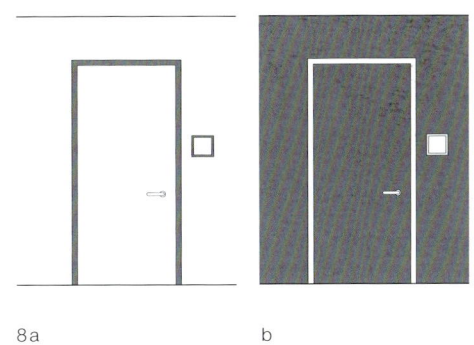

Cognitive impairments
Impairments to cognitive skills in the areas of perception, recognition, remembering, thinking, deductive reasoning and judging can essentially be divided into the following areas:
- Memory disorders
- Learning difficulties, forgetfulness, concentration problems
- Reduced abstraction skills and spatial imagery
- Personality changes
- Loss of social skills
- Susceptibility to disruptions

Designers working with facilities specifically for those with cognitive impairments should pay particular attention to the need for easy, self-explanatory orientation and hence a "legible" interior layout, i.e. additional clarity in the layout of facilities, which is crucial for supervisory personnel. The conscious use of light, colour and materials can help to improve orientation and also alleviate fears because boundaries are well defined, which increases the feeling of personal safety and well-being. Private rooms and places help persons whose memories and learning abilities have suffered, e.g. Alzheimer's patients, to find appropriate ties (Figs. 7a,b and 8a,b).

Mental impairments
Mental impairments, in a very similar way to cognitive impairments, express themselves in substantial, disorder-related behavioural or experience-type deviations, especially in the areas of thinking, feelings or actions.
However, the causes of mental illness are much more diverse and much more difficult to verify. A medically founded diagnosis with classic indicators is often very difficult to make. The effects on design can be compared with those for cognitive impairments.

Controls and handles
Automatic machines of all kinds and automatically operated elements on or in a building are becoming increasingly common in our everyday lives. Whether cash dispensers, vending machines or machines for issuing train or swimming pool admission tickets, there is still great room for improvement. Such machines must be designed in such a way that they can be operated by all people without restrictions – regardless of the degree of their mobility or sensorial impairments. Suitable measures must ensure that the machines are recognisable in the case of sensorial impairments, that they are accessible in the case of motoric impairments, and fully usable in both cases.

It is not only automatic machines that need to be considered; automatic doors must be looked at closely to ensure that they do not present any barriers, since it has become standard for doors to public buildings to open and close automatically. This automatic function must be readily evident, especially in the case of a side-hung door where a certain clearance is necessary to permit opening, or when the door can only be opened and closed automatically. This applies to the positioning of a suitable operating button which, when pressed, activates the automatic mechanism; quite obviously, it must be possible for anybody to operate this button, even those with motoric impairments. On doors that can only be opened and closed manually, it is often the case that the handles are too high for comfortable use by wheelchair users. Actuating the handle can be a problem for weak persons or those with motoric impairments when excessive effort is required. The height of the centre-line of any operating handle should always be 85 cm above floor level. In individual, justified instances, other dimensions in

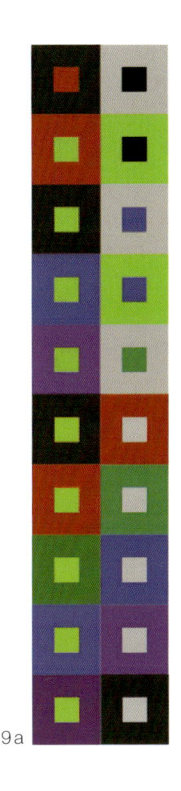

Design
Controls and handles

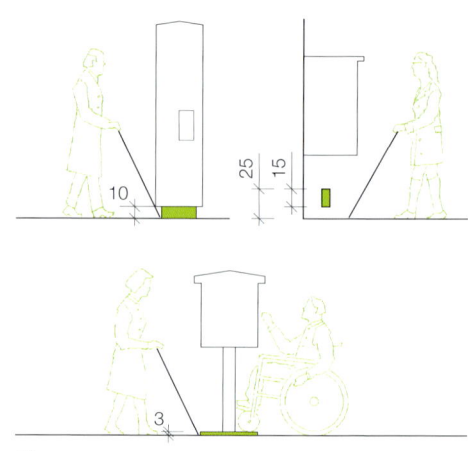

the range 85–105 cm are permissible (Figs. 10–12). The distance between the actuating element and any lateral constraint such as a wall, furniture or balustrade must be at least 50 cm.
The next step is overcoming the resistance of a door closer. Consequently, an optimum solution that satisfies all three conditions, i.e. distinguishability, accessibility and usability, is the automatic door, which uses a motion detector to control the opening and closing functions.

Other examples are automatically controlled systems and devices, the operation of which is not obvious, e.g. automatic machines controlled by smart cards. The slots for smart cards are frequently positioned in such a way that a person with impaired vision can hardly identify them, especially when there are other slot-type openings on the same device. In addition, their height above floor level is frequently inconvenient for wheelchair users or children, and they are impossible to use for that relatively large group of people who find it difficult to coordinate their hand movements reliably, e.g. following a stroke. A concave shell around the slot for the card, positioned between 85 and 105 cm above floor level, acts like a funnel and makes it easier to insert the card. This is an example of universal design; it fulfils the conditions required by the great majority of people.
Supplementary information regarding use is frequently included in the form of visual explanations or tactile information in Braille. The design of products, organised, planned and built environments, information and services should be based on the concept of usability for all persons equally. In this respect, no additional adaptation for special conditions, e.g. sensorial or motoric impairments, should be necessary. Communications equipment and technical devices forming a permanent part of the functionality of a building must be analysed with this in mind. Non-discriminatory utilisation avoids the exclusion or stigmatisation of persons with different capabilities: in lifts, for example, a horizontal control panel that can be operated from the sitting position, too. In public facilities, a toilet that can be approached from both sides, which allows for personal preferences and use from the left or right side, is another example of practical flexibility. Universal design criteria also take into account the fact that individuals move and work at different speeds. For example, a light control in a toilet should not be linked to a time switch.

Prompts and responses during the execution of a controlled process must always be readily understandable and unambiguous. With that as a criterion, a touchscreen for operating an automatic machine, for example, is then only acceptable when the actuation of a key is signalled by an audible signal. On the other hand, telephone keypads are designed to signal the actuation of a key once it has been depressed by a certain amount. Such designs apply the principle of two senses directly. First of all, there must be a clear view of just a few controls and their unambiguous manual accessibility must be guaranteed. The information required must be available irrespective of the surroundings or the sensorial abilities of the user; important information must be presented in visual, acoustic and tactile forms. Examples here are the information options in a lift: visually impaired users sense the functions of the buttons via Braille characters, and the visual indication is behind this transparent tactile cover; in addition, a loudspeaker provides supplementary acoustic information.
Installations with maximum fault tolerance can help persons with cognitive impairments. Haphazard or unintentional actions in such cases therefore involve only minimal risks or negative consequences. But the emergency call button on the lift-up support rail adjacent to a toilet cannot be fault-tolerant – and a corresponding warning sign must be provided to prevent accidental operation.
In addition to cognitive aspects, physical strains – or rather the relief thereof – also play a critical role in the design and positioning of controls. Efficient and convenient operation should be possible without fatiguing users unduly and without them having to adopt an unnatural posture. For example, handrails at the right height help to reduce physical strains. Ideally, operation should be possible irrespective of the size, posture or mobility of the user – a demanding requirement that cannot always be fully satisfied. Different hand and grip sizes should also be considered.

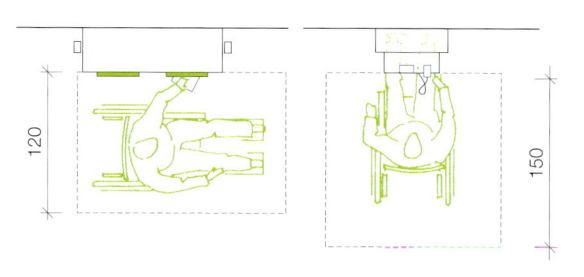

40

Design
Fire protection without barriers

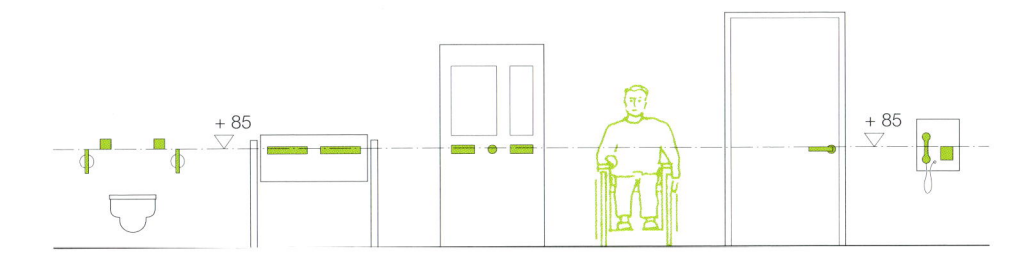

12

The total catalogue of requirements regarding universal usability has resulted in many detailed discussions over the years. Critics see the boundaries at points where there is a conflict of aims, or where a product, or rather a designed environment, can never offer a solution that meets all individual demands. In the preamble to the UN's "Convention on the Rights of Persons with Disabilities" it states that universal design does not exclude aids for certain groups of people with disabilities, where such aids are necessary. Simplicity and intuitive usability are among the most important aspects of universal design which, at the same time, however, are the most difficult to integrate fully.

Fire protection without barriers
The subject of fire protection in conjunction with disabled people is most certainly likely to cause conflicts and is the subject of heated debate among architects, fire safety specialists and representatives from interest groups. In principle, the requirement is that a building should be able to be used equally by all persons. But almost all buildings are not designed and constructed for the exclusive use of a certain user group and its specific requirements. One exception is perhaps a workshop for disabled persons. Buildings must permit flexible usage; a house, for example, should "grow old" with its occupants and at the same time remain suitable for a family. This demand is becoming evident in the current discussions surrounding demographic developments: the ageing tenants of an existing multi-occupancy building, for example, have only a narrow staircase and no lift, or at best only an excessively narrow one. The expression "A building only 'lives' when it is lived in" is often heard in this context. Basically, demands must be satisfied by the designers so that disabled persons can enter the buildings they design, use them, and, in the event of a fire, leave them – all without the help of others.
The fire scenario must take into account that lack of oxygen and the build-up of thick, frequently toxic, smoke represent a much greater risk than the flames themselves. According to the principle of equality, all persons, irrespective of any disability, must be able to save themselves by suitable means. Generally, there is a need to overcome a difference in levels.
In this respect, using ramps to supplement the stairs between floors is only realistic in exceptional circumstances because of the maximum gradient of 6 cm per 100 cm horizontal distance. Likewise, the requirement to equip all multi-storey buildings with lifts that can also be used in the event of a fire is unrealistic. Such lifts would have to comply with the highly complex, highly expensive attributes of the fire brigade lifts that are prescribed by legislation for high-rise buildings.
The first escape route required by building legislation, in the course of which changes in level are overcome by using stairs, is unsuitable for use by disabled persons without help. This also applies to the second escape route, again a mandatory requirement, which requires persons to use ladders – even for trained persons not the simplest of exercises and much more difficult for elderly or disabled persons. This problem inevitably results in a need to develop tailored solutions. One possibility for disabled persons is to provide protected waiting areas to which those with mobility impairments can escape and wait in safety until the arrival of the emergency services. Of course, from the psychological viewpoint, this is easier said than done. Another fire protection concept provides for authorised persons. Here, a person, a non-constructional measure, becomes the critical factor. So in addition to constructional fire protection measures, organisational measures – the establishment of precautionary measures for saving lives, the compilation of fire safety regulations and an alarm plan plus escape and rescue plans – becomes part of the concept. This is certainly only practicable when both regular training and corresponding fire drills take place.
Standard lifts may not be used for escape

T1: Design principles for building without barriers

Distinguishability	Clear view
	Appropriate marking
	Principle of two senses: access to visual, acoustic, tactile information
Accessibility	Right height: 85–105 cm
	Clearance to items at the sides: 50 cm
	Operability not dependent on size, posture, mobility
	Different hand and grip sizes taken into account
Usability	Fault-tolerant systems
	Minimum effort required
	Speed adapted to the individual
	Simplicity and intuitive usability

10 Vending and similar machines mounted on max. 10 cm high, narrower plinth or above tactile bar or 3 cm high "silhouette"
11 Movement areas for using machines either sideways or frontally
12 Mounting height for controls and handles

Design
Streets, paths, open spaces

13

in the event of a fire, although persons attempting to escape may choose them as a last resort in the face of no alternatives. In light of this, even standard lifts, e.g. in public buildings, have in a few cases been approved as a means of escape when their functionality can be guaranteed at least during the early stages of a fire. The supply of electricity to lift machinery must branch off from the main supply to the building in order to remain unaffected should the power supply to the building fail during a fire. The cables must be protected by a fire-resistant casing offering 30 minutes protection. A "close doors" button in the lift must override the photoelectric barrier that would normally close the lift doors automatically once significant amounts of smoke have been detected. Only with such an override is it possible to operate the lift at all in such a situation. These two measures are regarded by some specialists as adequate during the critical phase, i.e. the first 30 minutes, of a fire. After that, other conditions apply because generally the fire brigade with professional rescue personnel and equipment will be on the scene by then. However, approving all lifts as a means of escape does not comply with current legislation.

Persons with sensorial impairments are frequently left out of the fire safety equation. In most cases fire alarms are exclusively acoustic signals; deaf people therefore lose valuable time between the sounding of the alarm, the point when they are made aware of it, and escape or rescue measures. Acoustic fire alarms must be accompanied by unmistakable flashing lights or other visual signals in all the rooms of the fire compartment affected. This means that, for example, in a concert hall a visual signal must be available in the toilets as well as in other areas of the building.

Streets, paths, open spaces
In the light of the demographic factors already outlined in the chapter "Developments in society" (pp. 29–33), the constant rise in life expectancy and the increasing number of severely disabled people, public spaces will have to satisfy increasingly sweeping duties and obligations. Urban planners, engineers and architects must therefore design open spaces so that older persons and those with disabilities can use these easily and independently. The goal is to design the built environment without barriers so that virtually all people can make use of facilities and amenities in the generally accepted way without particular difficulties and certainly without having to rely on the help of others. Ultimately, these measures also benefit many other people, e.g. children, parents with prams and pushchairs. Public and private interests must be weighed up and considered when planning public streets, paths, open spaces, traffic infrastructures, parks, gardens and playgrounds. Private demands on public spaces can conflict with planning proposals. When considering the interests of persons with disabilities resulting from various capabilities or rather impairments, the conflicts between individual user groups force clients, i.e. local authorities, to accept many compromises and one-off design solutions. For example, planning proposals for facilities without thresholds are of great benefit to wheelchair users. But at the same time, fewer thresholds reduce the orientation options for blind people, for instance.
A series of acknowledged and legally binding directives or recommendations obliges local and regional governments to provide and maintain streets, paths and open spaces in accordance with all the requirements of safety and order. Recognised codes of practice, standards and directives are the tools for applying the measures to achieve freedom from barriers according to the relevant legislation. For example, article 3 of the Basic Law for the Federal Republic of Germany (*Grundgesetz*) calls for no person to be disadvantaged because of disability. The UN Convention, ratified almost 60 years later, goes even further, and speaks of "inclusion" (see "Regulatory principles", p. 17).
In the end, however, decisions have to be made regarding the application of these and other regulations in each individual case.

Often the objective of "no barriers" fails to meet the needs of all groups of people. Completely flat, non-slip surfaces without borders and changes in level are regarded as ideal for wheelchair users and persons with mobility impairments, whereas blind people need unambiguous, readily distinguishable information in order to move around safely. The 3 cm high change in level between different areas is generally acceptable for blind people, because they can still sense this, but just low enough for wheelchair users to negotiate without trouble. But even such a minimal change in level is often a problem for persons with wheeled walkers. Anybody who has suffered a sporting accident, for example, can readily imagine how quickly one can become a member of an affected group. Specific measures in public spaces therefore create "freedom from barriers" for larger groups of the population.
This expression initially encompasses only "accessibility for wheelchair users", but should and does include all types of impairments.
As previously stated, it should be possible for all persons to use the built environment and access public transport without any difficulties. This in turn means that users should be able to move around essentially without the help of others. This independence applies to the following

42

Design
Streets, paths, open spaces

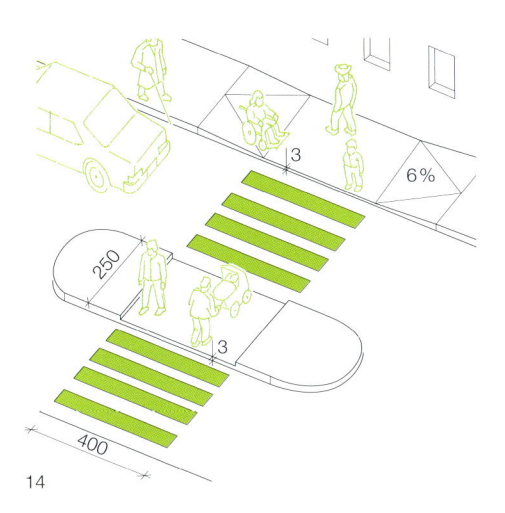

groups especially:
- Persons with motoric impairments
- Persons with mobility impairments
- Wheelchair users, also those with upper body disablement
- Persons who rely on mobility aids (e.g. wheeled walker, crutches, walking stick)
- Persons with cognitive impairments
- Visually impaired persons, i.e. blind people and those with impaired vision
- Hearing impaired persons, i.e. deaf people and those with impaired hearing
- Persons with other impairments e.g. rheumatism, diabetes)
- Older people
- Children
- Exceptionally short/tall people
- Persons with prams, pushchairs or luggage

Movement areas in public spaces
Freedom from barriers embraces the areas required in addition to the typical circulation areas. We should assume that persons, for instance, in wheelchairs, or with prams, or on crutches, or simply requiring more space because of luggage, could encounter one another here. Space for changes of direction must be available, too. Various dimensional recommendations are available for public spaces. For example, a pedestrian refuge between two carriageways should measure 400 × 250 cm, or 300 × 200 cm at pedestrian crossings (Fig. 14). Adequate space for wheelchair users is the reason behind these dimensions. In many situations a change of direction of more than 90° is necessary, and sufficient space for manoeuvring a wheelchair is essential. Such movements often require more space and so the movement area is defined as a square measuring 150 × 150 cm. A square of this size will be required as a refuge and also at the following places:
- To permit a change of direction
- At the start and end of a ramp
- In front of entrances, checkouts, security barriers, etc.
- In front of controls
- In front of and adjacent to benches
- In front of vending machines, letterboxes, intercoms, etc.

Passing places in public spaces
A quick change of direction with a wheelchair or wheeled walker is normally possible on a movement area of 200 × 250 cm. In practice, however, such generous dimensions are often not achieved because of a lack of space on the whole. It is always important to check how two-way traffic could be avoided. Traffic signs must be provided on both sides of narrow areas in order to warn users, and the signs should preferably indicate an alternative, barrier-free route.

Surface characteristics
The grip of a surface underfoot is critical for its use by persons, including those in wheelchairs, during any type of weather. The smoother the surface, the easier it is to negotiate with a wheelchair. But at the same time, it must also be rough enough to guarantee a non-slip surface for other users. Likewise, joints should be reduced to the minimum and should be filled flush with the surface. Proper workmanship is essential at changes of surface, and when renewing and repairing surfaces, so that no trip hazards ensue.
Surfaces with different tactile characteristics represent a valuable aid to orientation because they generate clear signals that can be felt and heard by a blind or visually impaired person using a white cane. For children, the elderly and visually impaired persons, it is important to provide a visual contrast between footways and cycle tracks. A 3 cm high continuous kerb between different usage zones is still widely used today. However, this does represent a hazard for cyclists and others. The preferred solution is to provide footways and cycle tracks in light/dark or contrasting colours and with a tactile difference between the two. The growing number of – essentially avoidable – collisions between pedestrians and cyclists in our busy towns and cities could be reduced by the appropriate constructional measures.

T2: Recommendations for guidance path surfaces inside and outside buildings [3] (see Fig. 17, p. 44)

	Dimensions	Dimension [mm] (tolerance ±0.5 mm)		
		Range	Recommended indoors	Recommended outdoors
a	Distance between crests of neighbouring bars	25–60	25–60	30–50
b	Bar width (in plane of measurement)	5–20	5–10	5–15
c	Clear distance between bars (in plane of measurement)	20–50	20–50	25–40
h	Bar height (trough to crest)	3–5	3–4	5

13 Change in the surface characteristics of a square in Rome (I), paved throughout with small stones, in order to make the main routes easier to use for wheelchair users, for instance, and at the same time offer tactile orientation aids.
14 Dimensions of a pedestrian crossing with pedestrian refuge

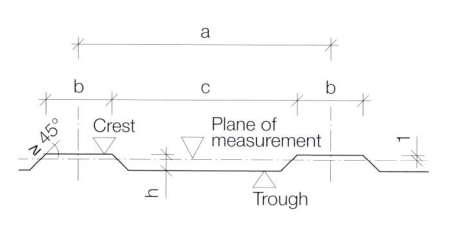

Design
Streets, paths, open spaces

15a

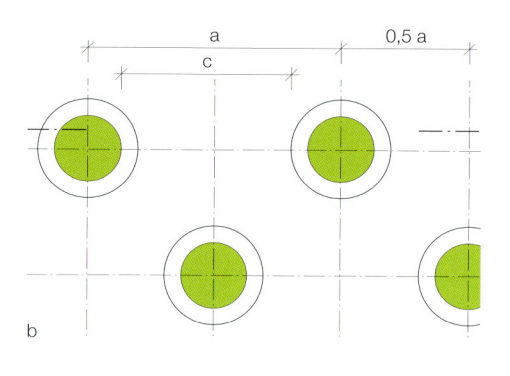

One obstacle for wheelchair users and those with mobility impairments is the paved or concreted channel between footway and carriageway, which not only acts as a drain, but also serves to denote the boundary between the two. The depth of this channel should not exceed the technical minimum dimension of max. 1/30 of the width because otherwise the channel will become a trip hazard.

For the blind and visually impaired especially, the material of the surface plays a very important role in their orientation. Tactile paving systems employ zones with contrasting tactile finishes to aid orientation (Figs. 16 and 17). Such "information surfaces" positioned in the course of footways must be provided at pedestrian crossings, escalators, lifts and public transport waiting zones to indicate the following:
- Changes of level (e.g. steps)
- Start, end or branching of guidance path surfaces
- Bus/tram stops to the side of the footway
- Railway level crossings
- Information systems for the blind and visually impaired

Information surfaces must be at least 90 cm deep (or at least 75 cm when using a vibrating metal system) and extend across the full usable width of the crossing or step. Where guidance path surfaces lead to an information surface, a square measuring 90 × 90 cm is adequate. Guidance path surfaces themselves consist of flat-topped bars that show the course of the route to be followed. The longitudinal profile (10–20 mm grooves) must always point in the direction of travel. Guidance path surfaces should be 25–60 cm wide [3]. In practice, however, these grooved elements with their sine-wave-shaped indentations have proved to be less useful than the readily perceptible, raised flat-topped bars or studs ("blisters") which, at the right spacing, can also be detected with the feet (Figs. 15a, b). Information surfaces should be positioned where persons with sensorial impairments are to be informed about a change of direction. Mobility training enables many visually impaired persons to gain a very good impression of the positions of important places and facilities. In addition, towns and cities are experimenting with different surfaces as aids for persons with sensorial impairments, to guide them to important buildings and amenities. Another possibility is to use signalling devices mounted in the surfaces of roads and paths, which use radio frequency identification (RFID) chips to send signals to a receiver in the white cane and not only guide persons, but also warn of particular hazards.

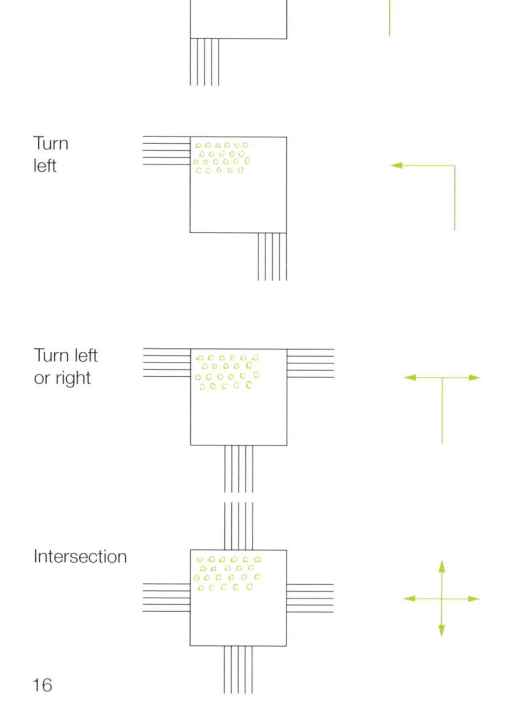

Turn right

Turn left

Turn left or right

Intersection

16

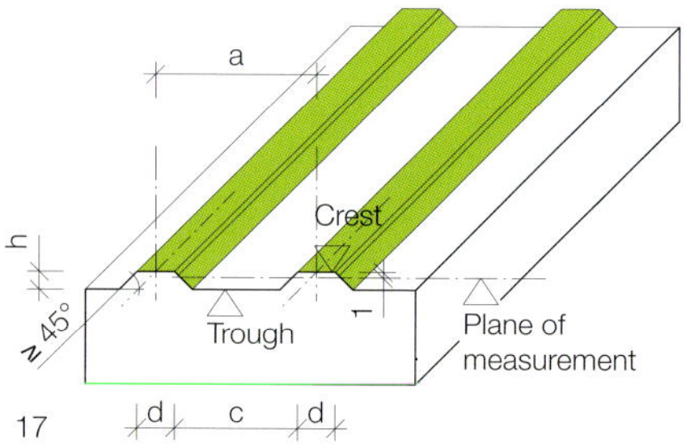

17

T3: Recommendations for information surfaces inside and outside buildings [3] (see Fig. 15)

Dimensions		Dimension [mm] (tolerance ±0.5 mm)		
		Range	Recommended indoors	Recommended outdoors
a	Distance between crests of neighbouring studs	50–75	50–75	50–75
b	Stud diameter (in plane of measurement)	15–30	15–30	15–30
c	Clear distance between studs (in plane of measurement)	min. 35	min. 35	min. 35
h	Stud height (trough to crest)	2–4	3–4	5

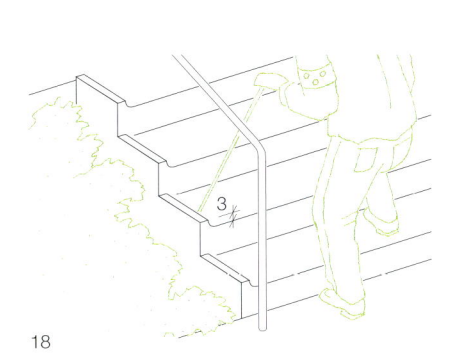

Design
Streets, paths, open spaces

Inclines and gradients
Differences in level represent special challenges for many people. Planning inclines and gradients in paths and open spaces therefore requires special consideration if they are not to present any barriers. A gradient exceeding just 3 % is already a problem for users of manual self-propelled wheelchairs, although a gradient of 6 % is regarded as an acceptable maximum for the entrance to a building. If steeper gradients are unavoidable because of the topography, suitable signposted alternatives must be available (Fig. 19).

Stairs and escalators
Fitting-out elements can also provide a sensible contribution to removing barriers. Those elements include lifts and escalators, orientation aids or bollards. Access to different levels is possible using various means. The handrail is especially important on stairs. The first and last steps of a flight of steps with more than three steps should be provided with a visual contrast – failing that, every step. The start and end of a flight of steps should be indicated clearly and in good time, e.g. by means of a tactile indicator on the handrails. It should also be possible to detect the end of the step with a white cane in order to improve the orientation for blind persons using the stairs. A uniform orientation system within a municipality, e.g. the same grooved paving elements, helps those affected. Information on which elements and forms should be used can be provided by associations for the blind. The consistent use of such systems is important.
Escalators are often too fast for disabled persons, which makes stepping onto and leaving the escalator very difficult. A maximum speed of 0.5 m/s is recommended. In addition, with a steep angle there is a risk of a loss of balance. The angle should therefore not exceed 30°.

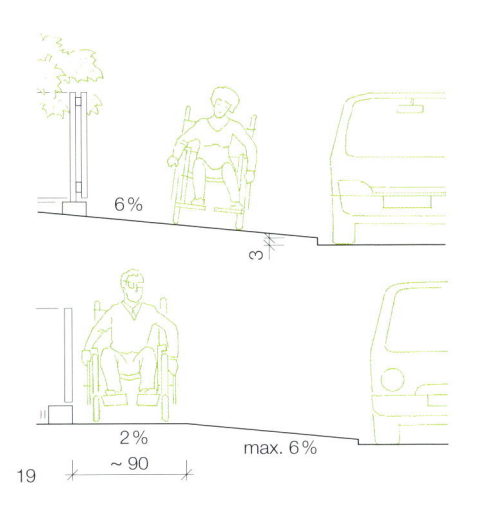

Parks, gardens, training areas and playgrounds
The popular and urgently needed leisure and recreational areas in towns and ities have been attracting more attention in recent years. These areas are intended not only to appeal to children, but also to older citizens, to offer them possibilities for exercise and communication. Differentiated design and usage concepts are important here. A few principles affect all target groups: usage without barriers begins with appropriate access. Orientation aids for the visually impaired in particular, e.g. tactile paving, low kerbs, colour contrasts, must be employed. Maps – possibly three-dimensional – showing footpaths, public transport stops, emergency telephones, toilets, etc. must be available. Emergency telephones help convey a feeling of safety and security.
Parks, playgrounds and other outdoor amenities often include safety barriers (which can also be in the form of dense planting) alongside paths on sloping ground. A gradient of 4 % is regarded as not unreasonable for outdoor facilities when there are no horizontal resting places in between. Wheelchair users and non-disabled persons should be able to sit next to one another on park benches to facilitate everyday communication.
When designing playgrounds, straightforward accessibility without barriers is important, but also adequate pathways and an infrastructure that includes public toilets. Playground apparatus must also be designed without barriers, i.e. must be suitable for use by all persons – an important social factor. The larger space requirements, i.e. movement areas for wheelchair users, must be taken into account. There must also be enough space to permit the switch from wheelchair to apparatus and vice versa. An area of 150 x 150 cm is necessary at changes of direction or waiting areas alongside apparatus. Nothing should project into this area to reduce its size.
Where movement areas for wheelchair users are raised above the surrounding area, then some form of kerb is necessary for a difference in levels of up to 15 cm, between 15 cm and 1 m posts plus a handrail are essential, and with a difference < 1 m a proper balustrade plus handrail. As accidents in the vicinity of playground apparatus are not uncommon, the designer must include orientation aids that warn, for example, the visually impaired against dangers such as swings and thus automatically safeguard the apparatus. The relevant dimension here is the potential fall height; the safety measures required depend on this dimension. All aids must work according to the principle of two senses.
Playground apparatus often requires a surrounding surfacing material that will cushion any falls. At fall heights of up to 60 cm the existing concrete or stone surfacing is often adequate, but at a height of 1.5 m a softer grass surface is essential. Around swings, but also the ever more popular climbing frames or walls, wood is a suitable material for cushioning falls. Wood chippings, bark mulch, sand and gravel are suitable for

15 Information surface: flat-topped studs ("blisters") [3]
   a Section
   b Plan
16 A square zone of tactile paving with flat-topped studs or a similar structure indicates changes of direction in a guidance path surface [3]
17 Tactile paving strip as a guidance path surface [3]
18 A 3 cm high kerb is helpful on steps that are open at the sides
19 Dimensions of transverse gradients on footways
20 Uniform illumination without glare or shadows is an aid to orientation and safety. Sergio Cardell tram stop, Alicante (E), 2007, Subarquitectura

Design
Streets, paths, open spaces

21

fall heights of up to 3 m, provided the layer of soft material is at least 20 cm deep. Synthetic materials can also be used to cushion falls. In addition, every piece of apparatus needs a clear area below and/or around it to prevent further injuries in the case of a fall. Where the fall height is 1.5 m or more, the width of this area must be at least equal to the height. The requirements become stricter as the fall height increases. The top end of a slide must be secured and slide users must come to a stop at the bottom end of the slide. Roundabouts must be designed in such a way that no thresholds have to be negotiated in order to reach them and they must be fitted with devices for securing wheelchairs against the centrifugal forces.

The discussion surrounding activity and mobility in old age is becoming increasingly important in what is actually a very heterogeneous generation of senior citizens and older adults. These members of society are active in walking groups, sports clubs and individual sports – activities that give them a certain self-affirmation.

But what is the situation outside organised leisure activities? The possibilities for older people are scarce and often all that remains is a walk in the park. However, numerous studies reveal that participating in sports activities is a very worthwhile pastime, especially for older people. Those who take regular physical exercise can prevent the onset of illness and also permanently enhance their well-being. The enjoyment gained through exercising with other people also helps to counteract loneliness in old age. Of course, sport does not arrest the ageing process, but it does have a favourable influence on the typical age-related changes. A lack of movement is a serious risk factor which leads to dependence on others and physical and mental limitations. Many people are basically dissatisfied with their lack of movement, but are not offered enough opportunities nearby.

A specific "playground" for older people could, however, serve as a training and communication zone to a certain extent. It is not performance that is important here, but rather simply movement and exercise.

Finland is one of the pioneers in this field. In a study carried out by the University of Lapland, a group of people aged 65 to 81 was encouraged to use climbing frames and ladders at public playgrounds. Afterwards, the group's suppleness, speed and coordination capabilities were much improved.

There are already several models for multi-generation exercising, i.e. for young and old; amenities with separate areas, where there is considerably less action than in the children's playground, represent another opportunity.

Such an amenity can be adapted to suit the needs of older people even more. Larger parks provide the chance of accommodating leisure activities for younger and older people within reach of their homes. The following activities are just some of the possibilities that can be incorporated in a training zone for senior citizens:
• Outdoor bowling
• Minigolf
• Outdoor chess
• Curling sheets
• Bocce or boules areas
• Basketball
• Beach volleyball court
• Table tennis table

So far, running tracks have been well used, and Nordic walking, sometimes on specially marked-out courses with obstacles and exercises, is enjoying a real boom. However, there is still a deficit in forms of movement that promote strength and skills. Innovative apparatus manufacturers have realised this and are now offering outdoor apparatus similar to those found in fitness studios, including special leg trainers or balance boards. Trying to throw a ball into a basket basketball-style but from a wobbling surface has been well received. Every piece of apparatus should include instructions regarding its use and purpose. In China public places include apparatus based on the principles of traditional Asiatic movement therapies such as Tai Chi and Qi Gong.

Various parts of the designated area can be reserved for Kneipp therapy pools and reflexology paths. As with children's playgrounds, apparatus and playing areas will need to be provided that take account of the needs of persons with disabilities. Another target group for such facilities could be the very old, who can now take part instead of just watching. Fragrance gardens or wheelchair courses are just two examples of many ideas for allowing older people to increase the quality of their lives. An appropriately designed "leisure oasis" could form part of the offers and activities of societies and educational establishments, e.g. Tai Chi and similar courses. Various conditions must be considered when positioning these amenities: adequate shade, good accessibility and sufficient space. The safety and security needs of older people is a subject that must be given due consideration. One balancing act in the design is combining good visibility of the area while at the same time avoiding putting the senior citizens "on show".

A certain degree of protection against vandalism and attacks is also necessary. Restaurants, cafés, etc. can promote the communicative aspect of these amenities, and a barbecue area is ideal for

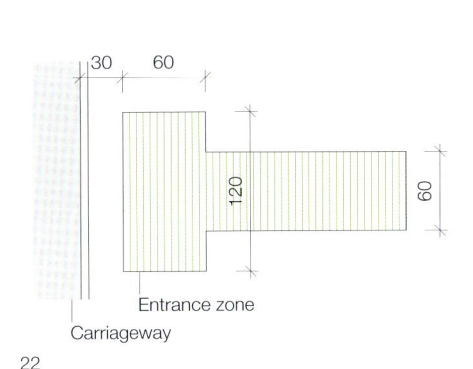

Design
Streets, paths, open spaces

organising individual events. A kiosk with a few tables might entice those who enjoy a game of cards. An existing restaurant with toilets would be an ideal starting point for setting up a new leisure facility for senior citizens.

Public transport
In technical terms, the use of public transport should no longer present any barriers because in the towns and cities, and increasingly in rural districts as well, all local buses make use of low-floor technology. This means that the design of the bus is such that the clearance between the floor and the road is minimised and the remaining difference in height is overcome by technical means. Hydraulic jacks, extending ramps and even the sideways tilting of the entire vehicle ease entry and exit.
Irrespective of problems caused by gaps and steps between waiting zones and the vehicles themselves, it must be assured that stops for buses, trams, etc. are fully accessible in the first place. Of the hundreds of railway stations in Germany, only a handful do not have steps leading to their platforms. Not only wheelchair users and the visually impaired have a problem with this; parents with prams and pushchairs, or cyclists, especially those carrying luggage, are also at a disadvantage. Funds for eliminating barriers from railway stations are often not made available until major urban projects are undertaken.
In order to improve customer orientation and improve levels of satisfaction, many public transport companies are now responding by converting their vehicles and stops into barrier-free zones. Disabled persons' equality legislation has been a major help in this respect. In principle, the goal is to make entering and leaving the bus easier, which in the end makes travelling by bus faster, safer and more economic, too. The gap between bus and footway must be given special attention. If the gap is too large, complications, even injuries, can be the result. A clearance of 3 cm is desirable, both vertically and horizontally. Innovative vehicle design systems allow the bus to "kneel", i.e. it can be lowered and tilted to one side. Many bus stops are too low, which means that the step up to the floor of the vehicle is too high. Raising the level of the ground and providing ramps can minimise the change in level. The approach surface should exhibit a gradient wherever possible but at the same time be smooth.
Further measures that can be implemented include guidance path surfaces for the blind and visually impaired. A tactile paving system, e.g. grooved panels, can direct such persons from the footway to the front entrance of the bus. An information surface should be installed up to the edge of the kerb in line with the front door of the bus (Figs. 22–25, p. 47).

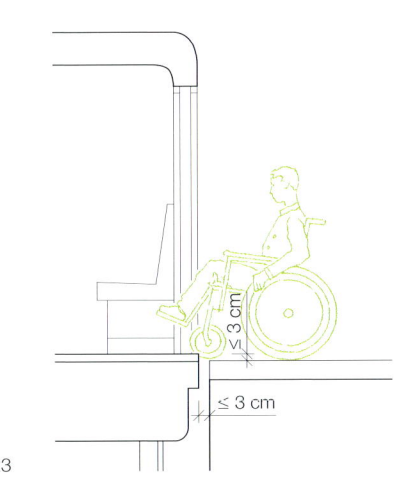

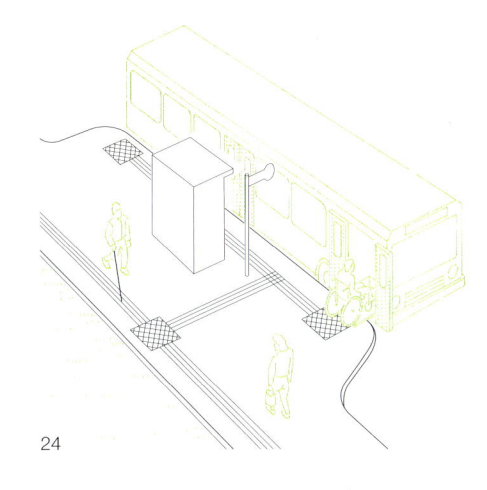

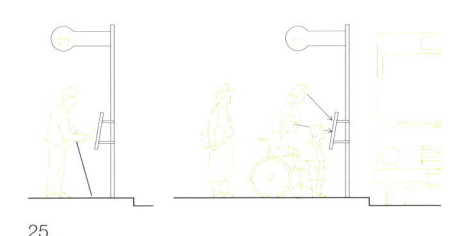

21 Glass doors separate the tracks from the platform and open simultaneously with the doors of the train as soon as it arrives at the station. Islands Brygge underground station, Copenhagen (DK), 2003, KHR arkitekter
22 Bar-type tactile paving to indicate the position of the doors for entering a means of transport [3]
23 Platforms and stops should be designed so that the difference in height and the gap between the interior of vehicle and the paving is no more than 3 cm.
24 Tactile paving at a bus stop
25 Information panel mounted at a height of 85–105 cm, with the information also available in Braille

47

Design
Streets, paths, open spaces

26

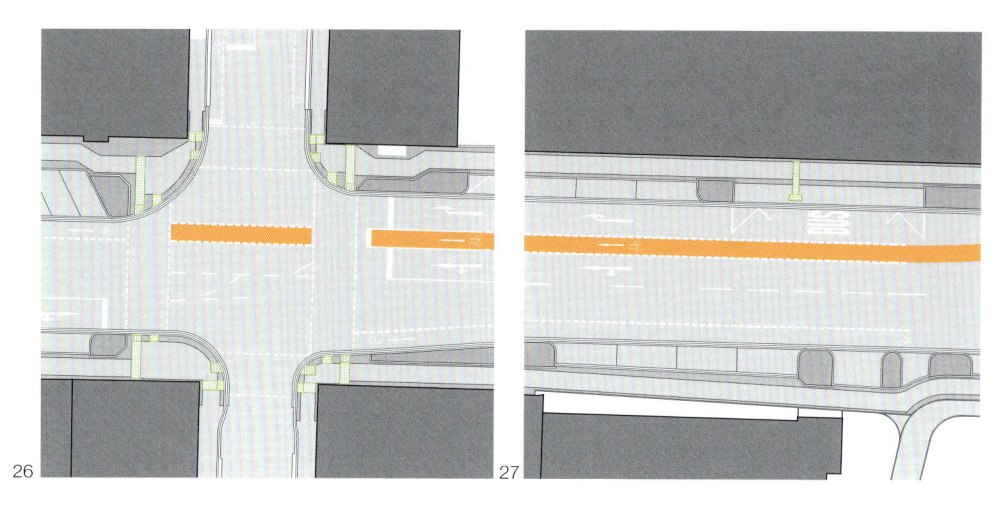

27

Pedestrian crossings
Crossing places with traffic signals are further important elements in designing without barriers within local communities. Conflicts will occur here due to the volume of traffic the road has to handle and the physical abilities of persons who have to cross the road when they have right of way, i.e. a green light (Fig. 14, p. 43). We assume a speed of 1 m/s for pedestrians, and this speed should be regarded as a minimum because of our motorised traffic flows. A central island in wider roads enables the green phase for pedestrians to be split. In such cases the planning team must weigh up very carefully who is to be given priority (vehicles or pedestrians).

The most diverse models evolve out of solving the conflict between the requirements of mobility impaired and blind persons: What is for one group a difficult-to-negotiate barrier, is for the other group an orientation aid important to safety. Besides the unsatisfactory compromise of the 3 cm high kerb, which is intended to satisfy the intrinsically contradictory requirements of blind persons and those with mobility impairments, there are two other aspects to be discussed: separate crossing places for different user groups and the modification of the common crossing place. This means that lowering the footway to road level is combined with tactile paving along the edge which warns of the dangers. The most important aspect may well be the consistency of the system; a local community should implement one solution throughout. That will also ease travelling for persons not familiar with the locality (Figs. 26–28).

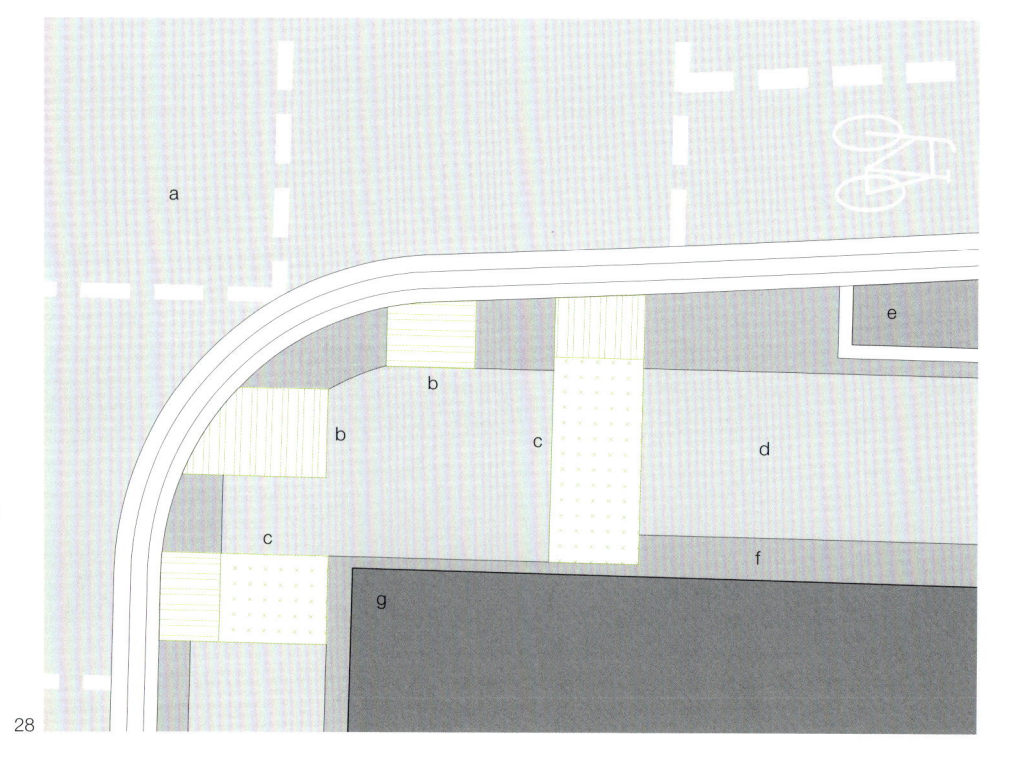

Barrier-free upgrade to Innstrasse in Rosenheim, Rosenheim Civil Engineering Department, Ing. Büro ROPLAN, Rosenheim
26 Combination of crossing with lowered kerb for persons with mobility impairments and crossing with kerb for visually impaired pedestrians
27 Bus stop
28 Detail of combined crossing
   a Road
   b Wheelchair users
   c Visually impaired
   d Footway
   e Planting
   f Cobblestones
   g Building

28

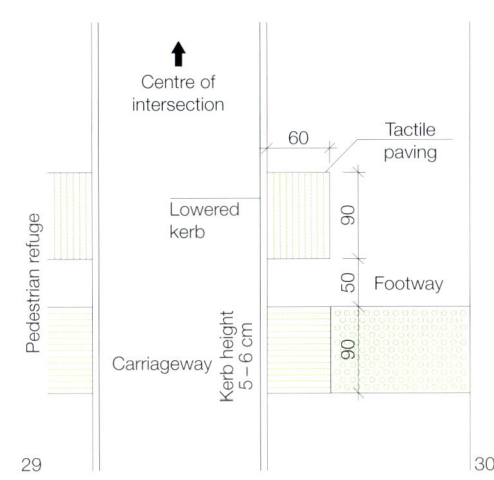

Design
Streets, paths, open spaces

Barrier-free parking spaces
As already mentioned in the example of parks and playgrounds, there is definitely a need to provide barrier-free access to public transport, public buildings and residential buildings.
Where private vehicles are in use, barrier-free parking spaces are also essential. Specific recommendations apply to car parks and parking spaces, especially in-line parking alongside roads. The latter should normally be at least 2 m wide and 5 m long; drivers enter and leave their vehicles on the road itself. Wheelchair users who drive themselves, as well as those using crutches, take longer to enter and leave their vehicles and are therefore exposed to the dangers of moving traffic for a longer time. Barrier-free in-line roadside parking should therefore only be provided on roads with a low traffic volume. One-way streets, on the other hand, allow a driver to enter or leave a vehicle on the left side, i.e. clear of moving traffic. Lowered kerbs make life easier for those on crutches, but a 3 cm kerb should be retained in order to maintain an orientation aid for the blind and visually impaired. On busy roads with traffic in both directions, barrier-free in-line roadside parking spaces should be lengthened to 7.5 m and widened accordingly in order to reduce the risks when entering or leaving a vehicle. A driver in a wheelchair can then reach the footway either in front or behind the vehicle (Fig. 31).
With head-on parking (i.e. at 90° to the carriageway), once again the movement area of 150 × 150 cm should be included, which results in a total parking width of 350 cm. Space can be saved by providing one movement area for every two adjacent parking spaces for disabled persons, which results in a width of 5.5 m for two parking spaces. However, this shared movement area cannot be guaranteed because it is impossible to predict every situation: Who needs the additional area: driver or passenger? Was the vehicle driven forwards or backwards into the parking space? Additional information (a sign designating the spaces is required anyway) could make drivers aware of the fact that the additional, manoeuvring space should be left clear between the two parked vehicles.
There should be a parking space for a minibus near each main entrance. Dimensions of 250 × 750 × 350 cm (height × length × width) are specified for this special vehicle, which may have a wheelchair access lift or ramp via a rear or side door.
Buildings accessible to the public are usually provided with a drop-off/pick-up zone so that visitors can leave/enter the vehicle directly in front of the entrance; while the visitors are in the building, the vehicle is parked further away.
Parking spaces for the vehicles of wheelchair users should be located in the vicinity of lifts in multi-level car parks above or below ground and in the vicinity of the main entrance for all other buildings. This is because wheelchair users, being lower down than other car park users, are less conspicuous and therefore particularly endangered by other vehicles using the car park.
Statutory instruments regulate the number of parking spaces to be provided, which depends on the usage of the facilities and the parking spaces concept of the local authority.

29 Schematic diagram of separate crossings [3]
30 Tactile paving system, Rome (I). The grooved panels continue only approx. 100 cm, provided no potential dangers or branches are to be expected. The kerb serves as a "natural" orientation aid in these transitory areas.
31 Dimensions of in-line roadside parking
32 Dimensions of barrier-free parking spaces

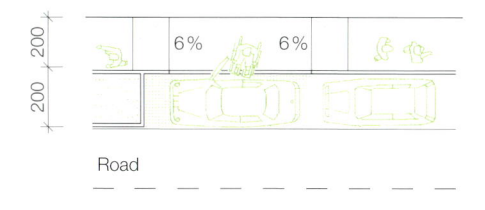

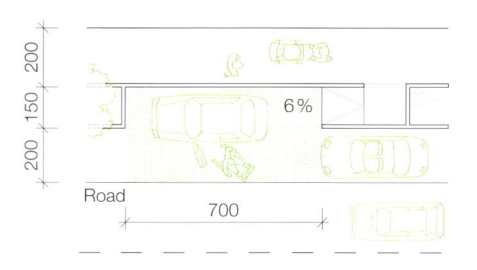

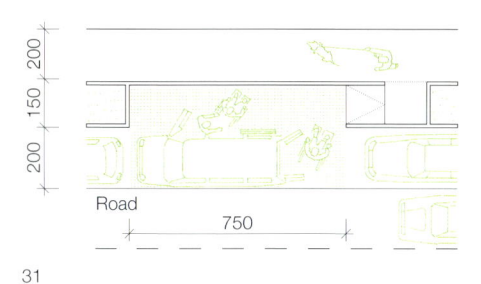

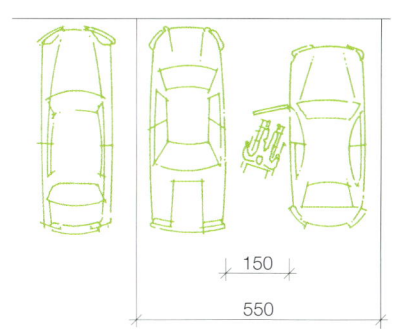

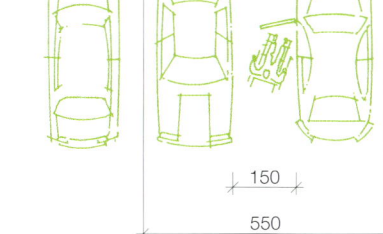

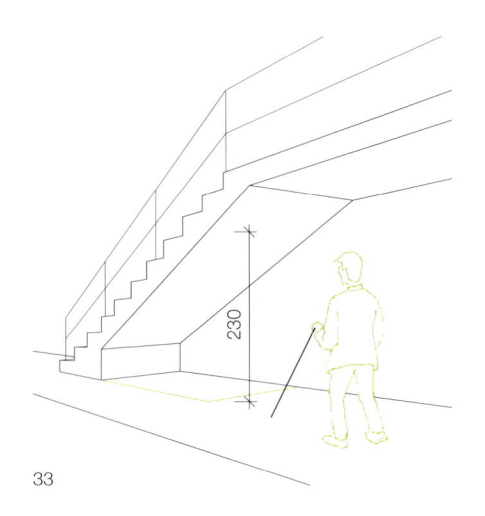

Design
Stepless accessibility, entrances and doors

33

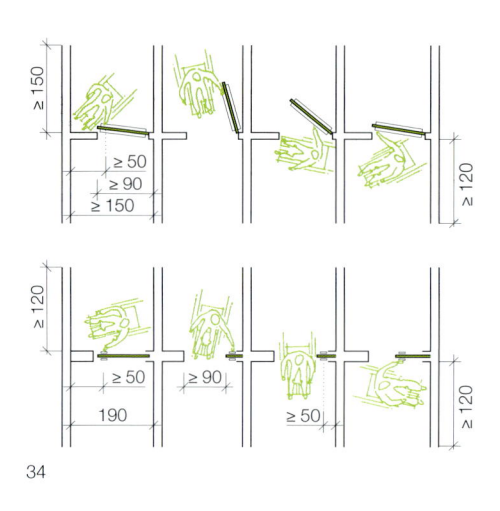

34

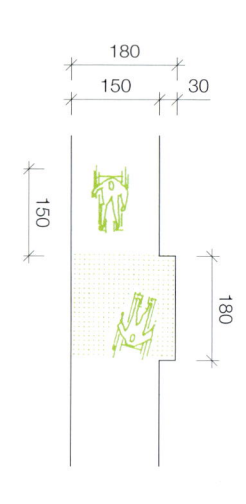

35

Stepless accessibility, entrances and doors

Leading an independent life strengthens our self-confidence, makes us less reliant on others. But to achieve such a goal, both in public and private life, constructional and technical prerequisites are necessary, which include the retrofitting of technical equipment in existing buildings. Above all, future design strategies must increasingly take into account the needs of older and disabled persons. It is the architect who first points the way towards a building without barriers: if the design embodies the social and statutory issues right from the start, then cost-effective, or at least cost-neutral, solutions can be found.

The route to the main entrance
An exemplary barrier-free fit-out to a building will be worthless if good access for the disabled is not guaranteed in the first place. The relationship between public spaces and the building entrance is therefore crucial. The external infrastructure must function in all weathers, at all times of the year.
The quality of outdoor footways leading to the main entrance is initially determined by their suitability for those reliant on wheelchairs or mobility aids. Besides an adequate width of 150 cm, there must be an area measuring 180 × 180 cm every 18 m at least to enable to wheelchair users etc. to pass. Disabled or older people can reach the limits of their strength when they have to negotiate gradients. The permissible gradient expressed as a percentage describes the ratio of the vertical distance to be overcome in cm per 100 cm of horizontal distance travelled; the longitudinal gradient may not exceed 3 %. A slope transverse to the path itself is often essential to ensure drainage of surface water; this may not exceed 2.5 %. For shorter paths < 10 m in length, a longitudinal gradient of max. 4 % is permissible. Surface finishes outdoors must be easy and safe for wheelchair users to negotiate, should not cause any undue vibration of chair or user. Wheelchairs require a firm, even surface.
Main paths, e.g. to building entrance and garage, must remain usable even during bad weather. Unmetalled surfaces and slabs or flags with a non-slip surface are ideal; gravel is obviously unsuitable for wheelchair users, likewise grass or "grasscrete". Paving stones can also represent a problem if the joints are too wide, and open-grid flooring is less suitable for older people and those reliant on mobility aids.
Some local authorities provide guidance systems as an aid to orientation outdoors. Such systems employ tactile paving zones supplemented by technical innovations such as satellite navigation via mobile telephone. Such systems are currently mainly restricted to the municipal infrastructure, e.g. tourist information, railway station, town hall. How each building is presented to potential users, how they are located, is the responsibility of the respective user or owner.

Residential buildings also have to satisfy requirements regarding access. The main entrance to a building should always be easy to find and must be accessible without having to negotiate steps or thresholds. A visually contrasting design for walls, floors, information systems, doors, doorbells and nameplates plus adequate lighting are vital for those with visual or cognitive disabilities. Tactile paving and floor finishes plus constructional elements that help to guide visitors, e.g. advertising signs, represent good orientation aids for the visually impaired.
Where the gradient of an access route to an entrance exceeds 3 or 4 %, lifts or additional, shallower ramps must be available. There must be a suitable, level movement area on both sides of an entrance door. Appropriate movement areas around doors are discussed below (see "Doors – geometry and hardware", also Tab. T4). Ideally, the main entrance to a public building or even a private residential building should be fitted with a canopy to provide protection from the rain.

Entrance doors
The entrance doors to public buildings are usually heavy. Often, disabled people can hardly muster enough strength to open them, which makes an automatic opening mechanism essential. In the event of a power failure, it must still be possible to open the doors mechanically or manually. Safeguards are necessary to prevent injuries if there is a risk of catching fingers or clothing. Rounded edges and rubber guards can reduce the risk of injury. The closing mechanism on an automatic door must be fitted with a safety device that stops the door if there is a risk of somebody being trapped. Contrasting markings on the floor can be employed to indicate the open position and hence warn of the dangers. Wheelchair users cannot use revolving doors, and such doors are also a problem for some old people and those with sensorial impairments. An alternative, side-hung door must therefore be provided alongside – a fire safety requirement anyway. Nevertheless, such an alternative entrance can lead to security risks in many instances, so this second entrance must be taken into account in the building's security concept.
It is not only non-automatic doors that need sufficient movement space on both sides; space in front of or alongside buttons for operating automatic doors is essential. Sufficient clearance between the operating button and the door leaf as it opens is not only obligatory, but simply common sense.

Design
Stepless accessibility, entrances and doors

A clear height of at least 230 cm (DIN 18040: 220 cm) is necessary to allow unobstructed passage. This dimension takes account of the fact that each subsequent generation is a little taller. The minimum clear width for entrance doors is 90 cm (Tab. T4).

Horizontal circulation
Steps or thresholds may not interrupt the circulation on one level. A single change in level > 2 cm makes access impossible for a wheelchair user. Users of mobility aids such as wheeled walkers or walking sticks are able to negotiate an intermediate step in some cases. However, no compromises should be accepted in the design of new buildings.
Corridors and other circulation zones must be designed with a sufficient width so that they are suitable for wheelchairs or mobility aids. A width of at least 150 cm is regarded as adequate. Additional areas must be provided so that wheelchairs can pass; corridors > 15 m long must include one area measuring at least 180 × 180 cm so that wheelchair users and persons with mobility aids can pass.

Handrails and display cabinets, for example, may not encroach on the recommended widths and heights. Individual components that do project into the usable floor space must also be readily visible to persons with impaired vision. In particular, components that restrict the area for movement, e.g. the underside of a flight of stairs, must be indicated where the available headroom is < 200 cm. This can be achieved by way of tactile building components or tactile flooring.

Doors – geometry and hardware
Doors are a complex design issue, not only because of the movement spaces they need to be allocated. Usage requirements and technical specifications demand a case-by-case design approach. For example, in buildings accessible to the public the specification for the movement area in front of lift doors is different to that in residential buildings, although the area must always measure at least 150 × 150 cm and may not overlap with other movement areas. This is especially important for busy lifts, e.g. in department stores, where congestion or two-way traffic could result

in obstructions and hazards. In residential buildings, on the other hand, waiting and circulation zones may overlap. Furthermore, the escape route regulations of the federal state building regulations also apply. A lift call button should preferably be mounted on the wall 100 cm to the side of the lift so that a waiting wheelchair user does not obstruct persons leaving the lift and is not forced to reverse.

Wheelchair users and those reliant on mobility aids are very limited in their movements. The movement areas on both sides of a door are of different sizes. On the side to which a side-hung door opens, the area should measure 150 × 150 cm because the wheelchair user has to move up to the door in order to actuate the handle and then reverse in order to allow the door to open (Fig. 34). When opening and passing through a door in the same direction as that in which the door opens, no extra manoeuvring is necessary, which means that the movement area need only be 120 cm deep. A transverse bar 85 cm above floor level

T4: Geometric requirements for doors according to the draft of DIN 18040 [1] [cm]

All doors		
Clear opening	Clear opening width × clear opening height above FFL	≥ 90 × 205
Actuating handle,	Clearance to building, equipment and fitting-out components	≥ 50
other handle, signage	Height above FFL	120–140
Manually operated doors		
Actuating handle	Height of axis of rotation above FFL	85
	Height of axis of handle/operating height always 85 cm above FFL; in justified individual cases other dimensions of 85–105 cm are possible	
Other handle, horizontal	Height of axis above FFL	85
Other handle, vertical	Height of handle above FFL	85
Automatic door systems		
Button	Height of centre of handle above FFL	85
Button for side-hung/sliding door approached sideways	Clearance to main mating edges	≥ 50
Button for side-hung door approached frontally	Clearance in direction of opening	≥ 250
	Clearance in direction of closing	≥ 150
Button for sliding door approached frontally	Clearance on both sides	≥ 150

[1] Part 1 has been approved and is due to be published in the late summer of 2010. Part 2 is due to be approved soon and should also be published in the late summer of 2010.

33 Space below flight of stairs
34 Sequence of movements for opening and passing through side-hung and sliding doors for a wheelchair user
35 Outdoor footways must include a passing place (180 × 180 cm) at least every 18 m

Brandhorst Museum, Munich (D), 2008, Sauerbruch Hutton
36 Main staircase
37 Grooves cut into the floor to form a tactile and visual warning of the underside of the main staircase

Design
Vertical access: stairs, ramps, lifts

38

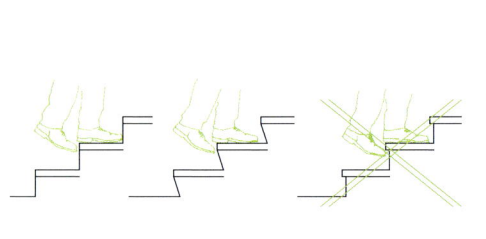

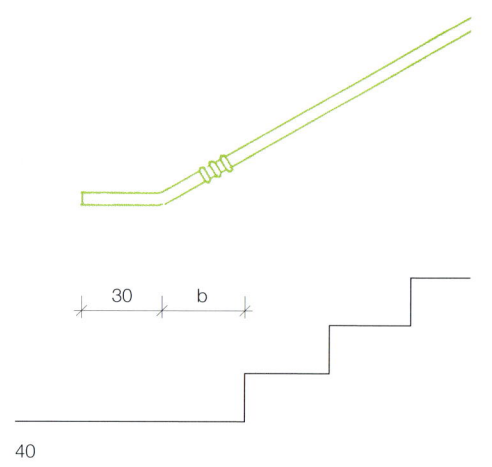

39

40

makes it easier for a wheelchair user to pull the door closed. A sliding door will require a movement area 120 cm deep and 190 cm wide on both sides because of the manoeuvring necessary to open and close the door (Fig. 34). Sliding doors must be planned very carefully because it is not the clear width between the jambs that is critical, but rather the clear opening between the edge of the open door leaf and the opposite jamb. The door handle must be easy to operate, from both sides, without any risk of injury, when the door is open. Recessed handles are unsuitable.

The depth of the jamb is also important: a door handle should be positioned no further than 25 cm back from the front edge of the jamb if it is to remain easy for wheelchair users to operate.

Only a small amount of effort should be necessary to open and close a door; where this is not the case, an automatic opening/closing system will be required. A delayed closing feature, either mechanical or electrical, will give persons with motoric impairments enough time to pass through the door safely.

Doors in public buildings and the entrance doors to dwellings must have a clear opening width of at least 90 cm. Barrier-free doors within dwellings may be just 80 cm, but they are then no longer suitable for wheelchairs. This is because although wheelchairs are rarely wider than 70 cm, experience has shown that a width of 90 cm is necessary to guarantee a safe, prompt passage (Tab. T4).

The opening angle of a side-hung door should be at least 90°, and doorstops, door handles or items of furniture should not reduce this angle. Depending on the type and position of the door hinges, the effective clear opening width can be reduced by the thickness of the door leaf,

and the designer should allow for this, especially in the case of thick door leaves. Basically, however, the clear opening width between the jambs is relevant. When choosing hardware, recessed, spring-loaded or other complicated fittings must be avoided at all costs because they make operation difficult, even impossible. The doors to toilets, showers and changing rooms may not open into those rooms. This is because such rooms usually comply with minimum dimensions and an inward-opening door would limit the movement options even further, rendering extra manoeuvring with a wheelchair necessary. In addition, in the event of accidents, which take place comparatively frequently in such rooms, a person lying on the floor could prevent the door from being opened or be injured as the door is opened. It must also be possible to open such doors from the outside.

Persons with impaired vision or cognitive impairments should also be able to find and identify doors without any problems. Contrasting colours, on floors as well, are just as important here as tactile floor finishes. A change of floor finish also provides information for blind people; a contrast here is understood to be a difference in the tactility. The use of design means, e.g. for advertising on the floor or advertising signs, must be carefully planned to avoid presenting the visually impaired with distracting signals.

User safety is almost more important than orientation. For example, glass or glazed doors must include safety markings to make them readily visible.

Vertical access: stairs, ramps, lifts
The stacking of levels one on top of the other and linking these by way of stairs, ramps or lifts is one of the outstanding inventions in the history of building. At the same time, all these vertical links are associated with problems, i.e. barriers. Persons with mobility impairments cannot use stairs. Ramps, owing to their large footprint, are normally only suitable for overcoming small vertical distances, not whole storeys. Lifts are ideal, apart from their comparatively high installation and maintenance costs plus their dependence on electricity.

Stairs
Stairs and barrier-free design would seem to be incompatible partners at first sight. This is certainly the case for those users for whom stairs represent an insurmountable obstacle, e.g. wheelchair users. Vertical access for such groups must therefore be guaranteed by way of alternatives. But for all other groups of people, suitable measures can help to create a "staircase without barriers", and convey a sense of safety and orientation. Barrier-free stairs must be designed very carefully in order to minimise the risks for persons using the stairs.

Handrails are crucial here. There should always be a handrail with a diameter of 30–45 mm mounted on each side of the stair at a height of 85 cm (top edge) above the pitch line of the stairs. The inner handrail around the stairwell may not be interrupted. Outer handrails must continue horizontally for a distance of 30 cm beyond the start and end of a flight of stairs, again at a height of 85 cm above floor level. This ensures that persons who are not quite so secure on their feet have already reached a landing or floor before they have to let go of the handrail (Figs. 38–40). This is why the inner handrail around a stairwell must be continuous: a disabled

Design
Vertical access: stairs, ramps, lifts

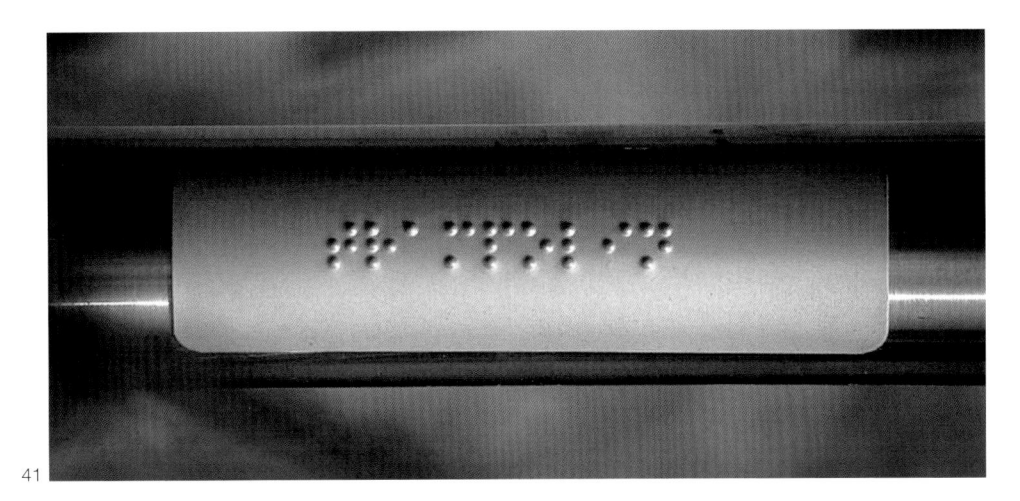

41

person should be able to rely on the feeling of security that a handrail provides, and not have to forego this even briefly. The inevitable crank in a handrail at every floor or landing is a signal to the visually impaired that they are approaching the start or end of a flight of stairs. Of course, the legal requirements regarding the height of a balustrade are unaffected by the handrail height of 85 cm.

A staircase should be illuminated in such a way that the nosing of the step does not cast a shadow, and the riser should be distinguished from the tread by using materials with surfaces of different brightness levels. Such details enable the difference between horizontal and vertical surfaces to be recognised more easily without irritating shadows, the contour of the edge of the step is clearer. Employing contrast in the design of the floor covering to a stair contributes to safety: when the floor coverings on steps and landings have different brightness levels and textures, the material on the landing at the bottom must extend as far as the first riser. At the top of the stair, the landing is already evident visually in order to express the change in level through the change in floor covering. As the final step is at landing level, this may not suggest any change in level.

Tactile markings on the handrails at the start and end of each flight of stairs can provide the visually impaired with information, e.g. storey number, escape route, etc. Apart from a number to denote the particular floor level, other information, e.g. about a particular room such as a canteen, can be helpful. Such tactile markings must be positioned on the outside of the handrail so that they can be felt with the tip of the index finger of the hand around the handrail (Fig. 41).

Compulsory stairs are those that form part of the first escape route according to building legislation. Their sizes, design and construction usually have to comply with certain requirements, irrespective of the size and use of the building. Germany's federal state building regulations normally specify a staircase enclosed on all sides when the stair serves more than three full storeys. It should be located on the external wall of the building, be fitted with windows and provide direct access to the outside. Compulsory stairs should be located no more than 35 m from any point in the building.

In principle, the design of staircases should be based on the use of the building and the anticipated number and abilities of its users:
- A lower level of fire protection is generally possible in buildings with a limited fire load.
- Where larger numbers of people congregate, e.g. in schools, places of assembly, department stores, the effective stair width of 80 cm normal in housing is not acceptable.
- If users with impaired mobility are expected (e.g. in hospitals, nurseries, homes for the aged), very high safety requirements apply to handrails, stair pitch and step design.

Compulsory stairs may not include any winders. Stairs with curving stringers, i.e. spiral stairs or free-form staircases, in prestigious buildings are permissible provided they do not form part of an escape route and provided there is an alternative vertical access route in the form of a lift in the immediate vicinity. A few basic rules must be observed: even if the depth of a tread remains constant along the walk line of a curving flight of stairs, the tapering form of the treads due to the curvature represents an uncertainty for the physically disabled and hence a potential risk. Certain one-sided impairments mean that it is only possible to use a handrail on the left or the right side. Stairs must therefore be easy to use throughout their width, a fact that does not apply to curving stairs or those with non-parallel stringers. At best, very wide flights and a minimal curvature to the inner stringer, i.e. a stairwell diameter of at least 200 cm, may be permissible, but the full depth of the tread must be available adjacent to the stairwell.

Risers contribute to a feeling of safety. With backlighting in particular, open-rise stairs or those with a very "transparent" design cause uncertainty, even promote a fear of heights. The right arrangement of tread and riser also helps to avoid accidents: the ankles and hips of the old, infirm or physically disabled are often less supple and there is a risk of catching the toe of a shoe when ascending. Projecting nosings, or rather set-back risers, are therefore not permitted. However, an angled riser is possible because it achieves a greater tread depth (Fig. 38, p. 52).

When planning an interior layout, it is important to ensure right from the start that lifts and open, descending stairs are not located opposite one another. Unclear situations, with many persons in or in front of the lift, haste or an enforced change of direction after leaving the lift do not need to be further aggravated by providing the risk of falling down an adjacent stair. If there is no alternative to positioning lift and stairs opposite one another, then a minimum distance of 300 cm between the two is essential.

38 Set-back stair risers of max. 2 cm are permissible.
39 Main stairs, Brandhorst Museum, Munich (D), 2008, Sauerbruch Hutton
40 Horizontal continuation of stair handrail
41 Handrail with label in Braille, patients garden, Großhadern Hospital, Munich (D), 2004, realgrün

Design
Vertical access: stairs, ramps, lifts

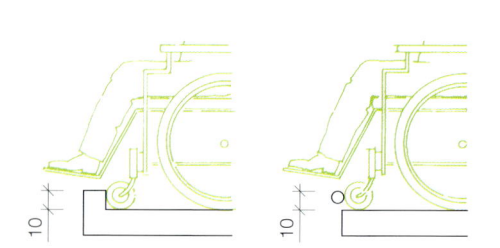

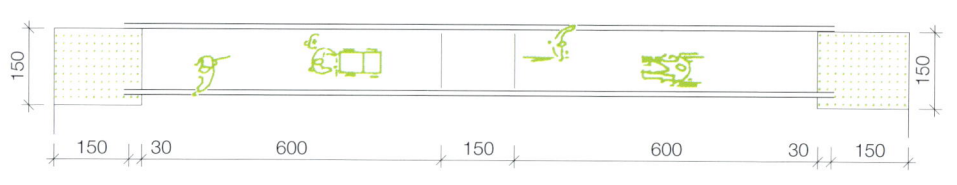

42
43

Ramps
Sloping surfaces can be used to overcome small differences in height. Circulation areas with an inclination of more than 3 % must be designated as ramps, which means that additional constructional measures are necessary. The gradient of a ramp may never exceed 6 %. However, circumstances that require a steeper gradient are frequently encountered in existing facilities. If a steeper ramp is unavoidable, this must be agreed between client, users and a disabled persons representative. The gradient must be assessed very critically: wheelchair users with an unstable trunk could tip forward when descending, and a dangerous feeling of weakness in certain susceptible persons when using a steep ramp cannot be ruled out. It is questionable whether the effort and time needed to negotiate a ramp between storeys really represents comfortable usability (Fig. 43).

The following principles must be allowed for in the design:
- Gradient: max. 6 %
- Min. usable width: 120 cm
- Intermediate landing 150 cm long on ramp lengths > 600 cm
- 10 cm high kerbs
- Handrails on both sides mounted at a height of 85 cm above floor level
- 150 × 150 cm movement areas at the start and end of every ramp

An intermediate landing at least 150 cm long is required to avoid exhaustion on ramps more than 600 cm long. This landing length is also adequate for ramps with two or more inclined sections. All ramps and intermediate landings must be provided with 10 cm high kerbs alongside open edges – where there are no upstands or walls – so that the small front wheels of wheelchairs cannot accidentally pass over the edge (Fig. 42). As with stairs, handrails 30–45 mm in diameter mounted at a height of 85 cm are required on both sides of ramps and intermediate landings. These are less important for wheelchair users of course, but essential for those on foot.

Lifts
A lift is a technical means for achieving barrier-free access to a building with more than one storey.
The lift car should have a minimum clear width of 110 cm and a minimum clear depth of 140 cm. This size provides enough room for one person in a wheelchair plus one or two other persons. In order that these persons can stand to the sides of the wheelchair, the doors should not be positioned in the middle of the car walls. A wheelchair user enters the car forwards and leaves it backwards, or possibly forwards in the case of a car with doors on two sides.
The floor size for a car with doors on two adjacent sides must allow for a 90° change of direction and therefore must measure at least 120 × 150 cm.
A handrail with a diameter of 35–45 mm must be provided on one side of the lift car; the same dimension applies to the clear distance between the handrail and the wall of the car. The recommended mounting height for the top of this handrail is again 85 cm. Horizontal lift controls can be fitted to the handrail, and positioning these at an angle will enable them to be seen and operated by both seated and standing persons. However, it is advisable to equip lifts as standard with vertical controls directly adjacent to the doors because blind lift users can find these easily without having to explore the entire car – which may be full of people – to find the horizontal controls. In this case a vertical control panel must include Braille labels.
The lift car should include a tip-up seat for older people or those with sensorial impairments, or balance problems, for

44

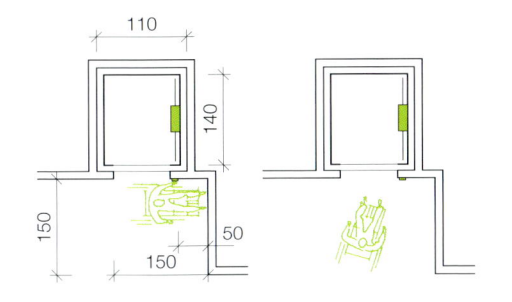

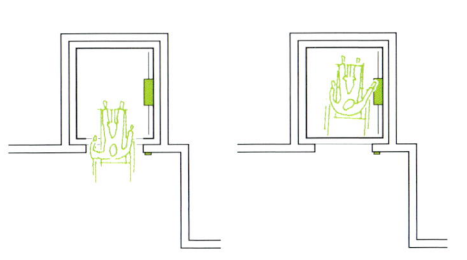

Design
Buildings and workplaces with public access

45

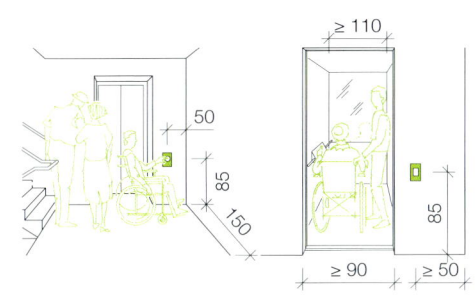

46

whom standing in a lift while it is moving might be an unnerving experience.
A lift car that does not permit a wheelchair user to turn around should be fitted with a mirror opposite the door so that the wheelchair user can see the space behind the wheelchair when leaving the lift (Figs. 45 and 46). Stainless steel has proved to be inexpensive and durable. Reflective metal surfaces on the doors are recommended for lifts with doors on both sides.

Buildings and workplaces with public access

Germany's Model Building Code (MBO) plus a number of federal state building regulations specify as a fundamental principle that all buildings and facilities accessible to the public plus a proportion of all new housing (irrespective of whether it is rented or owner-occupied accommodation) must be designed, constructed and maintained to meet the needs of persons with disabilities, old persons and persons with small children.

Buildings accessible to the public
Cultural events, eating out and sport (whether as spectators or participants) are all intrinsic to public life. All persons in our society must therefore be guaranteed access to, and full use of, all amenities. The goal of universal design begins in the entrance zone, which not only provides access, but also has to fulfil certain functions. Space for parking wheelchairs is one of those functions, which will require an area at least 180 cm wide and 150 cm deep, plus a manoeuvring area of the same size in front. It is here that many wheelchair users change from their everyday wheelchair to a different one that they use for work or sport. In swimming pools, too, spaces are required for special wheelchairs that enable disabled swimmers to reach the water.

The Model Building Code (MBO) calls for accessibility without barriers and appropriate use of those parts of a constructed, publicly accessible facility that can be used by visitors (cl. 50, para. 2). This applies to the erection, modification, change of use, disposal, usage and maintenance of such facilities. The MBO contains the following non-exhaustive list of typical examples of constructed facilities that – including their associated vehicle parking spaces and garages – must satisfy the above requirement:
· Cultural and educational facilities
· Sports and leisure facilities
· Healthcare facilities
· Office, governmental and court buildings
· Retail premises, restaurants
· Parking places, garages and WC facilities

Places of assembly
Rooms with rows of seats must include spaces for wheelchair users and their

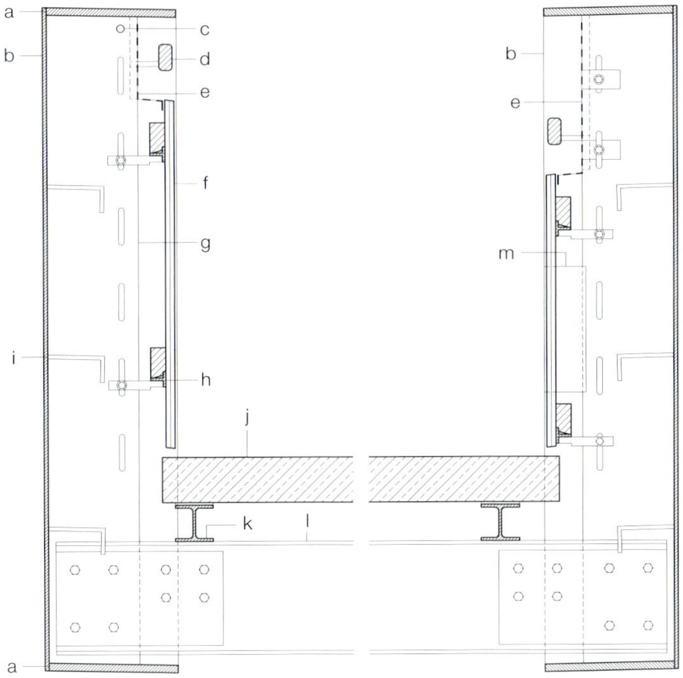

42 Kerb along the edge of a ramp without a side wall or safety barrier
43 Dimensions of ramps with max. 6 % gradient and intermediate level landings every max. 6 m

Römersteinbruch rock quarry, St. Margarethen (A), 2008, AllesWirdGut
44 Access to outdoor venue area via ramps
47 Section, scale 1:20
   a Capping, 20 mm pre-weathered sheet steel
   b Loadbearing framework, 10 mm pre-weathered sheet steel
   c Cold cathode fluorescent lamp (mountain side)
   d Handrail, 35 × 70 mm larch moulding
   e Perforated sheet metal screen, powder-coated
   f Larch planks, 24 mm
   g Stiffener, 10 mm pre-weathered sheet steel stiffener
   h Steel T-section, 40 × 40 × 5 mm, fixed to g by means of lug
   i Longitudinal stiffener, 150 × 70 × 10 mm steel angle
   j Precast concrete element, 120 mm
   k HEA steel section, 100 mm
   l IPE steel section, 300 mm
   m Built-in light fitting (valley side)

45 Plan showing minimum dimensions for lifts
46 Height recommendations and dimensions for lifts

47

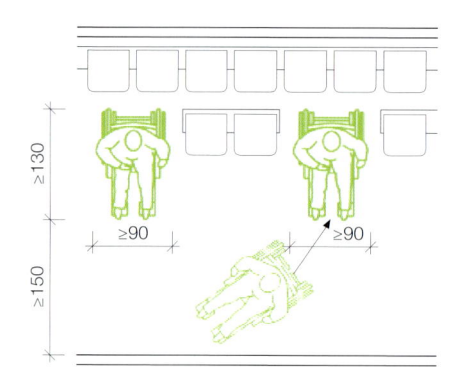

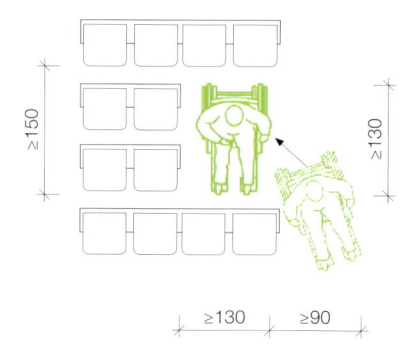

Design
Buildings and workplaces with public access

48

accompanying persons. The nature of the approach to such a space determines the size of the space and the associated manoeuvring area. Approaching forwards or backwards calls for different measures to an approach from the side. But in both situations movement areas may overlap with general circulation areas. The width of the parking space is 90 cm in both cases, the "net" depth 130 cm. We speak of "net" depth because the wheelchair fills this space. The overall depth is increased by a minimal aisle width between wheelchair and adjacent seats (Fig. 48).
The area 150 cm deep required to manoeuvre into and out of the space must be added to the "net" depth when approaching forwards or backwards. An approach from the side requires a clear area 90 cm wide. This dimension is similar to the normal aisle width required for a wheelchair. At least 1 % of all seats, but not less than two seats, should be designed like this. This figure is an approximation: in practice it has been found that the number of seats allocated according to this rule tends to be too few in facilities with < 1000 seats, but tends to be too generous for facilities with > 10 000 seats. The actual requirement is heavily dependent on the location and nature of the event. For example, in health spas and bathing resorts or at events dealing with the subject of disabilities in some way, an above-average number of guests in wheelchairs can be expected. Event organisers and hall operators must be able to allow for additional spaces or organise them at short notice. Removable seats allow the number of spaces for wheelchair users to be adapted to suit requirements. Where wheelchair users are accompanied by non-disabled persons, e.g. in concert halls, wheelchair spaces and normal seats must be combined alongside each other. An alternating arrangement of two wheelchair spaces and two seats in one row has proved effective in practice. Seats with more legroom are also recommended for persons with mobility disabilities and larger people. In rooms with fixed tables, e.g. lecture theatres, the spaces for wheelchair users must also be fitted with tables.
The areas provided for the audience/spectators should generally ensure a reasonable view of the performance/presentation area. This requirement is not always easy to fulfil: it is often the case in venues with tiered seating that only certain areas offer barrier-free access. Such areas are ideally located in the front third of the auditorium in order to meet the needs of older guests with poorer vision. In some large venues, screens provide a close-up view of the event for those sitting further back in the auditorium – a solution that can also be transferred to smaller events.
Aids for persons with sensorial impairments must be available in places of assembly as well as training and seminar rooms to ensure that they receive all the necessary information. The position reserved for a sign language interpreter should be readily visible from all places in the auditorium and equipped with separate illumination which must not cause any distraction when the room is darkened for using a beamer, for instance. Reading and writing surfaces for the visually impaired must also be illuminated. And where PA systems are in use, then there should also be an audio induction loop for those with impaired hearing. An audio induction loop enables those with hearing aids to switch into the frequency of the PA system. The system does not necessarily have to cover the entire room. In existing buildings in particular, e.g. old churches, only certain areas can be included for technical reasons. Suitable information signs are essential.

Design
Buildings and workplaces with public access

Sanitary facilities
Barrier-free toilet facilities, or even worse, the absence thereof, represents a challenge for any visitor to a town, restaurant or place of interest. Frequently located in the basement, such amenities are reduced to an absolute minimum size, which limits their usefulness for those reliant on mobility aids or those who require assistance in order to use the toilet. Such a situation is much more difficult for wheelchair users. Truly well-designed facilities are still rare. Keeping such facilities clean is especially important because direct contact with the toilet seat, for example, is unavoidable for wheelchair users. In public facilities this represents a problem where the frequent inspections and cleaning necessary to ensure reasonable conditions are not possible or it is necessary to pay a charge to ensure the presence of full-time personnel for cleaning and preventing vandalism.

One self-help measure that those affected can make use of is the pan-European network of public toilets that are only accessible to persons with a so-called Euro-key. These keys are available from various disabled persons organisations upon payment of a fee. One disadvantage of this solution, however, is that non-disabled older people are unable to have such a key. Another system frequently encountered is to deposit the key for the disabled persons toilet at a central point, e.g. with the gatekeeper, who then hands it out as required. The disadvantage of this system is that using the toilet is associated with an additional action, which has led to criticism. When planning public toilets or sanitary facilities accessible to the public, it is wise to obtain information about the existence and conditions of such options beforehand.

Comprehensive deliberations begin with the door: it must open outwards so that anybody who falls against the door on the inside cannot prevent it from being opened. That in turn requires a lock that can still be opened from the outside even when locked from the inside. Another positive aspect is that an outward-opening door does not restrict the space within the room. The clear opening width should be 90 cm and the door handle should be mounted at a height of 85 cm. A horizontal bar makes it easier for users to pull the door closed.

Appropriately sized manoeuvring areas probably contribute more than most other factors to the usability of a barrier-free toilet. The clear area in front of the WC, also in front of wash-basin and door, must be 150 × 150 cm. Further approach areas 90 cm wide and 70 cm deep are necessary to the left and right of the toilet (pp. 58–59, Figs. 51–54).

A disabled persons toilet should include the following elements: a combination of infinitely adjustable support rails on both sides of the toilet plus a backrest. It must be possible to fold the rails back out of the way from any position and they must be designed for a load of 1 kN applied at the end of each rail. The clear distance between the rails should be 65–70 cm and the top of the rails must be 28 cm above seat level. The reason for this relative dimension is twofold: it accommodates building tolerances and also enables an optimum force transfer irrespective of the absolute height of the toilet seat. The rails must project 15 cm beyond the front edge of the WC, a dimension that has proved favourable for the transfer from wheelchair to toilet. These dimensions are acknowledged as ergonomic and reduce the risk of accidents and injuries.

A toilet approachable from either side is standard in public facilities because user preferences and needs vary. It is the transfer from wheelchair to toilet that is primarily important here. This action can vary considerably between users because

48 Examples of wheelchair positions in auditoria
49 Sanitary space suitable for wheelchair user
50 Toilet suitable for wheelchair user

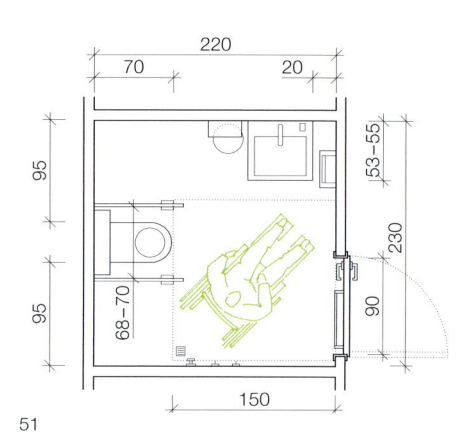

Design
Buildings and workplaces with public access

all have individual and diverse impairments to their mobility. The user grabs the support rails in order to move from wheelchair to WC, which in the case of a sideways transfer is positioned parallel to the toilet. The depth of the wheelchair or the distance from its front edge to the wall behind determines the depth of the movement area (70 cm). This is achieved either by using a special toilet that projects a sufficient distance from the wall, or by using a standard toilet that is attached to a false wall containing the plumbing. The size of such a false wall is critical: it must fit between the support rails and increase the distance between the front edge of the WC and the actual wall to 70 cm.

A backrest is required by those people who cannot support themselves properly while seated. Alternatively, the aforementioned false wall can help to keep the upper body upright. In both cases the distance to the front edge of the WC should be 55 cm.

The height of the toilet, i.e. the distance between top edge of seat and finished floor level, must be between 46 and 48 cm, a dimension that matches the seat height of the majority of wheelchairs. Higher seats are awkward, indeed dangerous, because it becomes more difficult to switch between wheelchair and toilet or to sit upright. In addition, it is easy for the feet to lose contact with the floor.

It must be possible to operate the toilet flushing mechanism from both sides with a hand or elbow without having to change position. There are two ways of achieving this: either an operating button on the wall behind the toilet, or electrical flushing with the switch integrated into the front end of the support rail. Each support rail must include a toilet roll holder within reach of the toilet user. Another position cannot be considered because many potential users have mobility impairments.

Proper use of the wash-basin is guaranteed in the first place by the height of the top edge, which must be 80 cm above the floor, and the provision of a clear space 90 cm wide underneath. Other designs, such as a recessed vanity unit with compartments on both sides and a "user bay" in the middle, are conceivable. The basin should be 55 cm deep overall. The form of the unobstructed movement area does not have to be 67 cm high over the full depth of the wash-basin.

Instead, we can assume a "stepped" arrangement because directly adjacent to the wall it is only necessary to provide space for the footrests of the wheelchair. It is the knee space that is critical, which can be guaranteed by fitting a suitable trap (Fig. 52).

If a wash-basin is fitted with a manually operated tap, then this must have a long lever designed for operation with one hand. Alternatively, a contactless system can be used, which is convenient, hygienic and saves water. Protection against scalding, i.e. a maximum water temperature of, normally, 38–45 °C (set by the manufacturer), is compulsory in both instances.

The mirror above the wash-basin – a standard fitting in any toilet – must be suitable for use by both seated and standing persons. Tilting mirrors are only permitted in exceptional circumstances owing to their poor appearance and the fact that every user first has to adjust the mirror. It is much better to fit a vertical mirror, the bottom edge of which is just above the top of the wash-basin (however, that calls for a pillar tap). A long mirror enables both tall and short, standing and seated persons to see themselves without difficulties.

51 Plan of barrier-free toilet room
52 Legroom requirement beneath a wash-basin
53, 54 Heights and clearances required in a barrier-free toilet room
55 Shower showing tip-up seat and dimensions of fittings
56 Shower area flush with surrounding bathroom floor

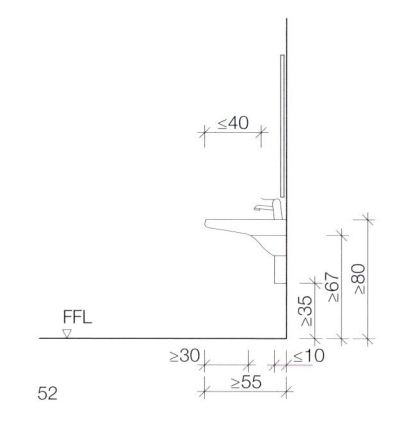

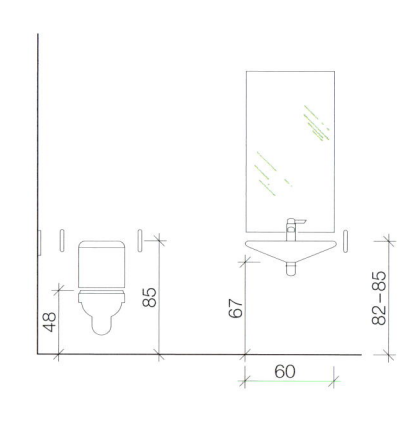

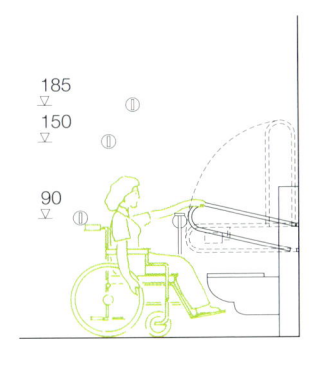

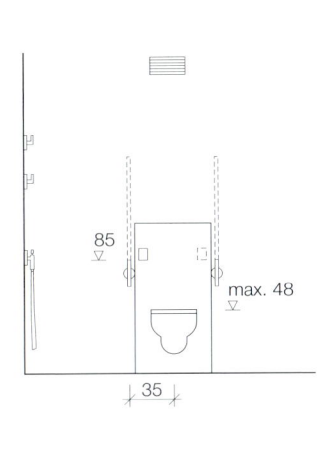

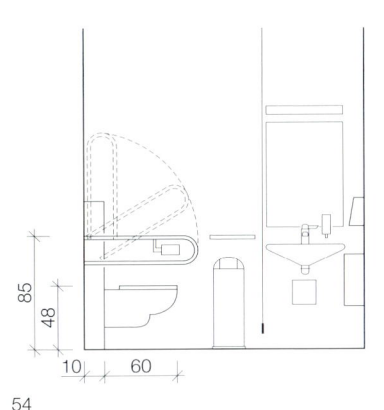

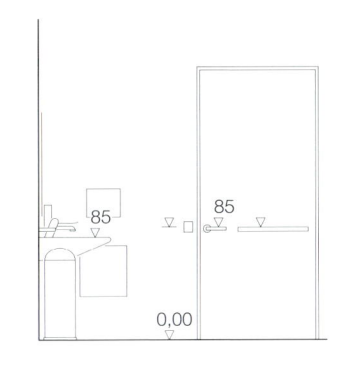

Design
Buildings and workplaces with public access

53

54

A manually operated soap dispenser must be fitted within reach, either above or adjacent to the wash-basin, likewise a paper towel dispenser or hand drier. These should all be fitted about 85 cm above floor level. The basket for used paper towels should be positioned below the paper towel dispenser if possible and should be a narrow model. The provision of a self-closing, 85 cm high, odour-tight waste receptacle for disposing of used sanitary articles is also recommended. It should be easily accessible and easy to use with one hand (Figs. 51–54).

There are further criteria that apply to the other technical fittings that enable usage without barriers.
Ventilation should preferably be mechanical. If there is a window, the handle must be mounted at a height of 85–105 cm so that it can be opened from a seated position as well. A motion sensor for controlling the lighting is always a useful addition to a light switch; this is because wheelchair users, for example, have to carry out a whole range of activities when visiting the bathroom. The activation time of the lighting must certainly be adjusted to the longer time that some disabled people will need when using the toilet.
An emergency call system is compulsory, and this must be operated by a pull cord fitted to the wall so that even a person lying on the floor can summon help. An emergency call button is also frequently integrated into the front end of a support rail, although one problem here is accidental actuation. It is also important to consider where best to direct the emergency call; reception staff or a caretaker would seem to be a good solution. But a toilet open 24/7 might need to be connected to a security service via a cellphone network.

The acknowledged codes of practice at first seem to leave little leeway for the number of barrier-free toilets to be provided in a building or one functional unit of a facility. Every sanitary facility must include one WC designed for disabled persons. In practice we find different interpretations of this rule; a disabled persons WC is frequently integrated into the standard sanitary facility, even if this is in a basement. The important factor is to apply a strategy for the location and number of toilets: near to a lift or a central position where possible are important criteria. Unisex or separate-sex usage is a discussion that always arises. Integrating one WC into each standard separate-sex facility is the optimum solution. In certain buildings that are not at risk of vandalism, e.g. office buildings, restaurants, having a WC that can be used by both disabled and non-disabled guests represents an economic solution.

Showers, which are found in the most diverse functional circumstances, must always be designed without any steps or thresholds, i.e. at the level of the surrounding floor, without a raised shower tray. A max. 2 cm deep tray in the floor is an alternative provided the 150 × 150 cm shower area does not overlap with the movement areas provided for other sanitary appliances. This area also permits rotation through 360°, and with a tiled floor throughout, this is a space-saving solution. The shower area can also serve as the clear area necessary to the side of the WC. A 40 cm wide and min. 45 cm deep tip-up seat with backrest, seat height 46–48 cm, must be fitted in the shower. Similar to requirements for the independent use of the toilet, a movement area 90 cm wide on both sides of the seat and 70 cm

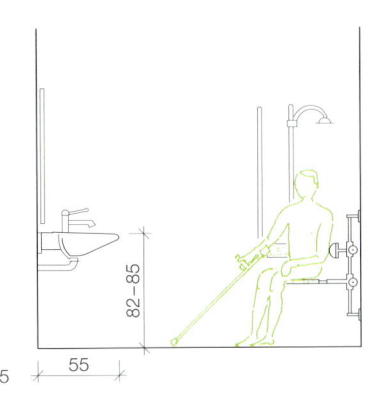

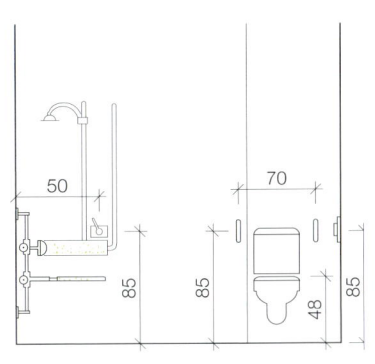

55

56

59

Design
Buildings and workplaces with public access

57

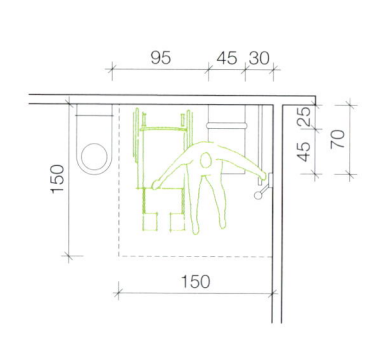

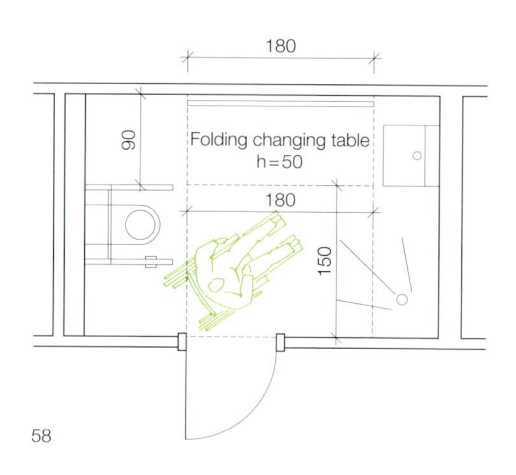

58

deep (measured from the front edge of the seat) must be available. Portable stools are very practical, but more suited to private or hotel situations because they are very difficult to secure. Fold-down support rails on both sides are also compulsory for showers. Horizontal and vertical grab rails also ensure an enhanced standard of safety and comfort (Fig. 55 and Fig. 57, p. 60).

If the shower is fitted with a fixed shower head, then there should be an additional, switchable hand-held shower head as well that enables a good wash from the seated position, too. The mechanism for fixing the shower head at the desired height must be considered very carefully: turning small knobs is difficult for persons with motoric impairments. A simple, self-explanatory and easily handled design is therefore more suitable. The same applies to the tap, which should also ensure that mixing hot and cold water does not result in a risk of scalding. A non-slip floor covering in and around the shower must be selected according to the current non-slip classes.

Clothes hooks at heights of 85 and 150 cm plus a separate shelf 15 cm deep and 30 cm wide at a height of 85 cm are helpful.

Barrier-free sanitary facilities as described above must also be available to the visually and hearing impaired. Even though these target groups do not necessarily need such large movement areas, grab rails, protection against scalding and emergency call buttons are useful features for these people as well. A high-contrast design for sanitary facilities helps persons with impaired vision to locate and use the sanitary appliances.

Sanitary facilities for disabled persons in roadside service areas should include at least one adult changing table – in combination with a shower. Where such a table is available, a baby changer is unnecessary because babies and infants can be changed on the adult table. Toilet, shower and changing provision in one room is popular with disabled travellers. A folding table is certainly possible in order to save space (Fig. 58).

Sports facilities
The importance of sport for older and disabled people should not be underestimated, both for the physical and the social/commutative aspects. Disabled persons use sports facilities such as gymnasia, swimming pools and fitness studios, either individually or as part of a team. Once again, the principle of general usability by all persons applies. A few specific measures in changing rooms and showers are necessary for persons with impaired mobility.

Correctly sized movement areas and an adult changing table are essential for the independent use of a changing room. The table, which is also used by persons who cannot use the lower half of their body for support, should measure 180 × 90 cm. A table narrower than 90 cm is not recommended because there is a risk of injury for persons using the table alone. The table mounting height of 46–48 cm corresponds to the seat height of a wheelchair. A movement area 150 cm deep must be available in front of the table.

The non-independent use of changing rooms, e.g. by severely disabled persons, must also be considered. In such a case the table is used for dressing and undressing, also for changing adult incontinence articles. It should be padded and fitted with a raised head cushion; a roll of paper for covering the surface of the table is advisable for reasons of hygiene.

59

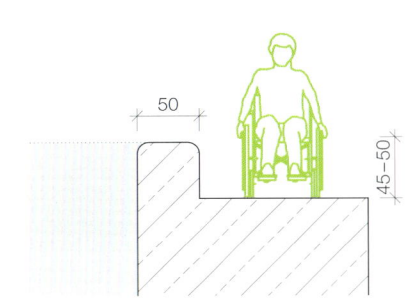

Design
Buildings and workplaces with public access

60

61

Swimming and therapy pools require technical aids, e.g. lifts, shallow steps, to help people enter and leave the water (Fig. 60). Spaces for parking wheelchairs must be provided depending on the chosen method of entering and leaving the pool. Ideally, special wheelchairs should be available. These are made of plastic and swimmers can take these right into the water; a visit to the sauna is also possible because there are no metal parts that can heat up dangerously. In many places the chairs on offer include beach wheelchairs, whose large wheels are also suitable for rolling over the edge of the swimming pool so that users can swim off directly from the seated position. However, their use presumes that personnel are available to assist the disabled bather.

For safety, visually impaired guests require contrast between the different components of the facility. Steps protruding into the water or ledges must be of a different colour, especially as the water makes it even harder to see such features.

Service and sales
To enable use by wheelchair users, the height of counters, ticket windows, sales displays, etc. should not exceed 80 cm. Where there are several identical units, at least one should be at this level and include clearance for knees underneath. As with wash-basins, room for knees and legs should be 55 cm deep at a height of at least 67 cm; once again, a stepped arrangement is possible.

Unfortunately, the standard designs for such fitting-out elements are less than ideal for wheelchair users or other per-

57 Movement area and approach to toilet in barrier-free toilet room with shower
58 Sanitary space in roadside service area or sports centre
59 Swimming pool "Le Bains de Docks", Le Havre (F), 2008, Jean Novel
60 Access to swimming pool via an upstand matching the height of a wheelchair seat; SIA, Switzerland
61 Spectator seating at Kohlbruck adventure pool, Passau (D), 2000, Teppert & Heiss, with Hiendl & Partner
62 Counter with legroom for wheelchair user in "gast", "Gasteig" Arts Centre, Munich (D), 2004, Zeeh Bahls & Partner Design, with Atelier Lups

62

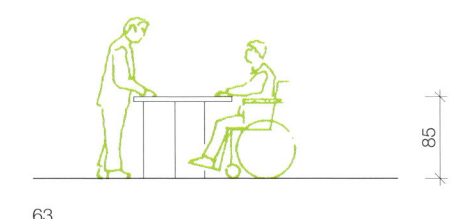

Design
Buildings and workplaces with public access

sons with impaired mobility. Floor surfaces at different levels in front of and behind a counter or ticket window can help to satisfy the requirement. Good signs and other visual means should be provided so that people with hearing impairments are able to receive important information and advice. Placing information at eye level is just as important as good legibility.

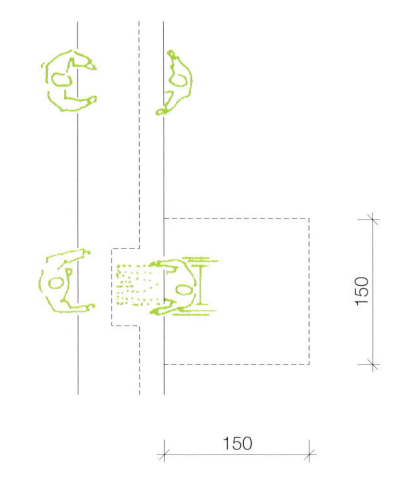

Places of accommodation
Offers from "barrier-free" holiday destinations read very differently and only very rarely is it possible to assess or even compare these. Standards and building regulations require barriers to be abolished in certain areas, but such requirements normally relate to new buildings and are not always fully implemented by smaller institutions. One very helpful tool in such cases is the development and adoption of a target agreement. Groups representing hoteliers and the like should cooperate with disabled persons organisations to draw up criteria that allow hotels and restaurants to be assessed according to minimum standards. It has become apparent that a classification can take into account different types of disability. What does not sound like an integrative measure at first is increasingly becoming a worthwhile option when classifying those existing buildings that owing to difficult framework conditions cannot eliminate all barriers for those with mobility impairments. In Germany operators have the chance, with the help of checklists, to assess their establishments themselves first of all, for compliance with minimum standards, and then apply for a classification. The German trade association responsible for hotels and restaurants can assign establishments to one of five categories:
A   No barriers for guests with a mobility impairment who may be partially reliant on a non-motorised wheelchair or walking aid.
B   No barriers for guests who are unable to walk and are permanently reliant on a wheelchair.
C   No barriers for guests who are visually impaired or blind.
D   No barriers for guests who are hearing impaired or deaf.
E   No barriers for any guests with physical or sensorial impairments.

Basically, the requirements to be satisfied by public buildings and housing have been transferred to the classification for hotels and restaurants.
For example, a hotel room with category B bathrooms corresponds to residential accommodation suitable for wheelchairs. On the other hand, category A does take account of barrier-free housing but is mainly characterised by smaller movement areas. The minimum standards of category B include the requirements of category A. In category C, contrasts play a major role for guests with impaired vision. The communication and alerting options in hotel rooms are important aspects in category D; for example, guests with hearing problems must be able to communicate by e-mail via an Internet connection in their rooms.
At least one single or double room generally complies with the above criteria.

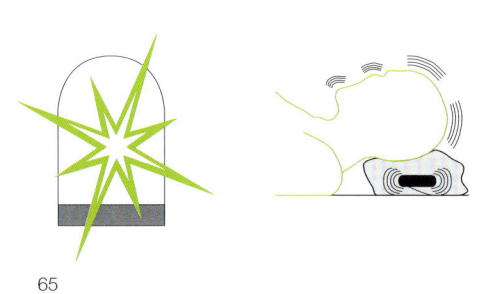

63 Movement areas and heights for counters and tables
64 Legroom beneath a counter over a width of 90 cm and adjacent movement area
65 Alerting options for guests with hearing impairments in places of accommodation; source: City of Graz
66 Furniture with legroom in seminar centre at Schloß Hohenkammer, Hohenkammer (D), 2007, HildundK
67 Furniture with recessed plinth and designed for approach from the side, Side Hotel, Hamburg (D), 2001, Jan Stormer and Mattheo Thun

Housing
The demographic changes in the majority of Western, industrialised nations and a paradigm change in lifestyle forms for persons with disabilities and older people mean that the policy-makers are having to face new issues for the housing sector. Urban planners, developers and architects are making "building without barriers" part of their itinerary. They are being forced to investigate in more depth the creation of barrier-free housing by new political terms of reference and models – also the new "UN Convention on the Rights of Persons with Disabilities" – plus social and market-oriented requirements.

Integrating disabled persons and old, non-disabled persons in a common living and social environment is the number one priority in housing policies. Barrier-free dwellings should be a matter-of-course standard for all housing. The advice centres of the chambers of architects and the cooperation of the relevant associations and organisations have helped to ensure that important findings are incorporated into the drafting of directives and legislation. The recommendations should be fundamental to the creation of housing so that more and more people with mobility impairments gain the chance for more independence, more mobility, which helps them to take part in their social environment (see "Developments in society", pp. 29–33).

A planner can systematically pursue individual goals. The key lies in the provision of clear, barrier-free horizontal and vertical access to all rooms and an appropriate internal layout. These principles are to be recommended for new-build projects and also the modernisation of existing buildings. The following considerations concern conversion work.

Firstly, it is important to check whether at least one level, in most instances the ground floor, can be reached without having to negotiate any barriers. This level can be redesigned as a whole or section by section.

If up until now the entrance to the building was reached via steps, a ramp or lift (possibly a stairlift) should be considered to overcome this barrier.

The first decisive step has been taken when the entrance doors to buildings and individual dwellings are readily accessible. Most entrance doors, apart from old buildings with double-leaf doors, comply with the minimum adequate clear opening width requirement.

The ideal – but expensive – solution is to install a lift that provides access to all levels. A lift with doors on opposite or two adjacent sides may be necessary in order to reach a raised ground floor. If a staircase is to remain the only means of access, a second handrail is absolutely essential, but this should not encroach on the width of an escape route as required by law. Prams and pushchairs are not the only items that now have to be parked in the staircase; increasingly, wheeled walkers must be left here, too. Access to a basement or underground parking is frequently complicated by the fact that these are often reached via steps. If the lift does not continue down to basement parking level, the recommendation is to create outdoor parking spaces for mobility impaired persons near the entrance to the building. Heavy doors, e.g. those to basement parking, can be retrofitted with automatic opening/closing systems. Dwellings are classed as barrier-free when they can be reached from public rights of way without having to negotiate steps, or via ramps or lifts. The entrance door to a residential building and a dwelling itself must have a clear opening width of 90 cm. Corridors leading to dwellings must have a clear width of at least 120 cm and include adequately sized manoeuvring areas in front of doors and lifts. There must be at least one manoeuvring area measuring 150 × 150 cm to enable wheelchair users to turn around. Within a dwelling, doors with a clear opening width of 80 cm guarantee usability.

The usability of a dwelling itself is just as important as accessibility of the dwelling. Adequately sized movement areas must be provided in hallways and all habitable rooms. In bathrooms, a shower flush with the floor is one of the key items. In this respect, the planner must check the technical feasibility in detail. The waste-water pipes from such showers can present a problem because the depth of the floor construction in existing buildings is frequently inadequate and there is also a risk of structure-borne sound transmissions. One possible solution is to equip ground-floor dwellings with floor-level showers because it is possible to route the waste-water pipes through the basement, below the ground floor slab.

When refurbishing bathrooms and adding pipework in front of existing walls (concealed behind dry lining), it is worth including reinforcing elements so that support and grab rails can be fitted if required. Usability without barriers is guaranteed provided the principal habitable rooms (living room, bedroom, toilet, bathroom, kitchen/kitchenette, room with washing machine connection) include adequately sized manoeuvring areas according to DIN 18025-2. Such areas must measure 120 × 120 cm. Appropriate movement areas in front of sanitary appliances (WC, wash-basin, bath) must be verified. Germany's Model Building Code (MBO) merely regulates the accessibility of buildings with more than two dwellings. However, a number of federal state building regulations now specify that in buildings with more than two dwellings, all the dwellings on one floor must comply with

Design
Housing

St. Cajetan, Munich (D), 2008, Ebe & Ebe
68 Covered entrance design with rest zones
69 Kitchen design for wheelchair users with the all-important sink across one corner

70 Kitchen island with legroom, Haus CK, Munich (D), 2002, lynx-architecture

barrier-free stipulations regarding accessibility and usability. This means that living rooms, bedrooms, one toilet, one bathroom, the kitchen and a room with a washing machine connection must have movement areas according to DIN 18025. In the meantime it has been established that the above requirement does not necessarily have to be satisfied by placing all such dwellings on one floor, but that compliance is also guaranteed provided the dwellings on several floors can be reached without having to negotiate barriers (e.g. by the voluntary installation of a lift not actually required by legislation). This enables, for instance, particular site conditions to be taken into account and several identical internal layouts to be built economically, one above the other.

According to MBO cl. 39, lifts are required in buildings in which the floor level of the highest storey containing habitable rooms is on average > 13 m above the surrounding ground level. Such lifts must be accessible (without steps and suitable for wheelchairs and wheeled stretchers) from all dwellings designed for barrier-free use and from all public rights of way. Where several lifts are installed, at least one should be capable of accommodating a pram, wheelchair, wheeled stretcher or goods, and this lift must stop at all floors.

Cl. 34, para.6, sent. 2 of the MBO specifies that stairs must be fitted with handrails on both sides and intermediate handrails where this is necessary for safety reasons. However, some federal state building regulations stipulate this for stairs in buildings with more than two dwellings reached via stairs. If stepless access (i.e. ramp or lift) to these dwellings is not possible, the use of the stairs by older occupants should be made safer and more comfortable by providing a handrail on both sides. It becomes clear from this that a detailed knowledge of the respective regional legislation is indispensable because critical stipulations relevant to the design may have been included in the clauses regarding building without barriers (see "Regulatory principles", pp. 20–21, Tab. T2).

Developers and architects should attempt to find out about the current and potential occupants of dwellings in advance. Large apartments at ground floor level with direct access to a garden appeal to families, but also to senior citizens looking for a new type of home. Design and construction strategies that allow apartments to be combined at a later date if required frequently lead to new utilisation options. For example, two smaller apartments could be combined to form one large, shared apartment for several senior citizens – with two bathrooms (essential in such a situation). The second kitchen could be converted for a different usage.

Housing cooperatives are increasingly planning one ground-floor apartment as a communal area in order to meet the needs of their ageing tenants. One large common room, with an additional WC alongside, takes on the character of a cafeteria and the occupants are provided with a common meeting place.

Exchanging apartments would seem to be an attractive proposition: where the mobility of a tenant is on the wane, swapping to an apartment with an identical layout but on the ground floor enables a person to remain within the same housing complex. However, this is dependent on many factors, not least of which is the popularity of such a move with the other tenants and the fact that a vacant apartment must be available at the right time. These aspects are also relevant to the combining of apartments. Such considerations must be taken into account prior to carrying out conversion work in order to clarify potential access issues and the coupling of corridors.

Barrier-free residential complexes consider not only restricted mobility, but also how our senses weaken as we grow older. Contrasting colours improve perception, a fact that is particularly important on stairs in order to reduce the risk of falls. Doorbells and letter-boxes must be labelled with larger fonts. Emergency or service call options are crucial to a better feeling of security and safety.

Users too – in consultation with the owner – can employ individual strategies in order to deal with "their own" barriers and gradually eliminate these. Considering the walls: Could any be removed in order to create more space? The demolition of a wall could turn a hallway into the movement area so useful to wheelchair users or those dependent on walking aids. Living and cooking areas are often in separate rooms in older buildings. If the intervening wall is removed, this opens up new possibilities for the ensuing open-plan interior. In older buildings WC and bathroom are frequently combined to form one room so that the extra space gained enables the sanitary facilities to be used independently for longer. A similar situation can be envisaged for a detached house: a cloakroom and a separate WC could be combined to form a barrier-free bathroom and turn the ground floor into a realistic option for living on one floor only.

Finally, the fittings themselves must be examined in a critical light. For example, the washing machine could be incorporated in the kitchen, below the worktop, which obviates the need to negotiate the stairs down to the basement every time, or creates more space in the bathroom. Access to a patio or balcony can sometimes be eased by way of simple measures: the combination of a small, possibly removable, ramp on the inside and a timber grating outside can help to overcome the door threshold.

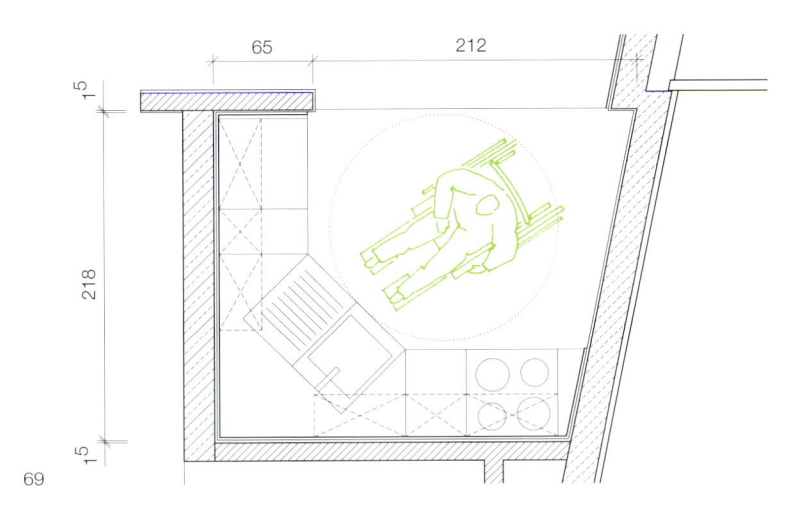

Hallways in dwellings
Efforts to design economical interior layouts has led to the minimisation of circulation areas in dwellings (so-called ancillary areas), especially where space is at a premium. This is a regrettable development, not only from the point of view of building without barriers: the hallway – the entrance and distribution zone in a dwelling – has become supplanted, replaced by a narrow, poorly proportioned, purely "functional" area without any habitable quality. The demand for a minimum manoeuvring area of 150 × 150 cm to allow a wheelchair user to turn around and turn through 90°, e.g. in order to approach a door leading off the hallway, envisages a certain spaciousness that has been missing in many dwelling layouts of recent years.

A clear width of 120 cm between walls is adequate for wheelchair users travelling straight ahead, without the need to turn. This is also the minimum dimension for persons with mobility impairments, who need extra space for their walking aids – a requirement that is often underestimated. Fittings and furnishings in entrance and hallway areas, e.g. coat racks, cupboards, may not encroach on the aforementioned dimensions. Consequently, this results in the need and the chance to re-evaluate and redesign the areas of dwellings that are wrongly classed as ancillary areas and thus neglected.

Living room, bedroom and kitchen
Barrier-free requirements do not enable us to define any absolute room sizes. Minimum sizes result from the footprints of the furnishings and fittings themselves, the distances between these regarded as adequate and necessary plus the indispensable area for movement within the room. Criteria for barrier-free usability of living spaces are the approach to, and operation of, windows and doors to the outside, also shelves, cupboards and seating. The switchover between wheelchair and seating must also be considered. Maximum generosity in the sizing of the areas for movement within the room which permit the trouble-free and effortless completion of all daily activities is another, equally important, aspect. This results in the need for a thorough examination of the fittings and furnishings required in dwellings. The general minimum requirement is an area measuring 150 × 150 cm for turning a wheelchair. Also important are manoeuvring areas in the bedroom, which is frequently too small. It should be possible to position a single bed so that it can be accessed from either side if necessary. Light switches and power sockets should be positioned to suit such a bed position. Besides the aforementioned accessibility of window handles, it must be possible to reach and use the bed and the cupboards and wardrobes usually found in a bedroom. The bedroom dimensions given in Fig. 72 (p. 66) are regarded as necessary for wheelchair users.

A wheelchair user requires an area 150 cm deep alongside the bed for manoeuvring and transferring to and from the bed. The space alongside the bed of a non-disabled partner should be 120 cm deep so that the wheelchair user can care for that person in the case of illness. In front of cupboards and wardrobes, a depth of 150 cm is required over the entire width so that they can be opened and used properly. Persons with mobility impairments but not reliant on a wheelchair require less space. In this case 90 cm is adequate in front of cupboards and wardrobes, and the space alongside a bed need be only 120 cm deep.

The size of a kitchen, as opposed to an open-plan living room/kitchen layout, results from the dimensions of the appliances in conjunction with the movement areas necessary for trouble-free use – even when it is necessary to work mainly or completely from the seated position. Positioning the most important items, e.g. cooker and sink, across the corners is an important consideration. Only in this way is it possible to avoid time-consuming and tiring travelling and manoeuvring. Positioning the main work surface in a corner enables cooker and sink to be reached with a minimal rotation of the upper body, without having to change the basic position very much – for wheelchair users and many older people a key prerequisite for using a kitchen independently (Figs. 69 and 70).

Height-adjustable work surfaces, a much-debated topic, would seem to be indispensable when 85 cm is chosen as the top level for appliances and surfaces. The exception to this would be alternate use by persons of very different heights. In recent years the manufacturers of kitchen units have specified much higher worktops because of the increase in the average height of the population. Individual planning and adjusting the height to suit is recommended.

Working in the seated position inevitably results in the need for knee space below work surfaces, also below cooker and sink. The cupboards so common below worktops can therefore be dispensed with. The maximum reach height from a seated position is about 140 cm above floor level; so wall cupboards for persons who must work from a seated position are virtually useless. Refrigerators, ovens and cupboards mounted at a height of 40–140 cm are recommended in kitchens for wheelchair users. This means that a barrier-free kitchen requires a much larger floor area, the size of which is not laid down, however. Neglecting these factors can quickly lead to considerable limitations in the functionality of the kitchen. Barrier-free appliances must comply with

Design
Housing

71

special requirements regarding their operation and control: the operating status must be visible and distinguishable from the seated position and for those with sensorial impairments. The operation and feel of controls for those with motoric and sensorial weaknesses must be considered, and that means many factors have to be considered in the design of such appliances. Drawers instead of deep compartments in low-level cupboards, and an oven with a pull-out carriage are features that ease the use of a kitchen.

Private bathrooms
As with the planning of other so-called ancillary areas (e.g. hallways), the design of private sanitary spaces is frequently based on minimum dimensions. The situation is particularly critical here because it is very difficult to compensate for a bathroom that cannot be properly used by a person with mobility deficiencies or a wheelchair user. It is frequently the bathroom that decides whether a person can remain in his or her dwelling after the onset of some disability. The cost of subsequent conversion and adaptation is often disproportionately high. Wheelchair users require the largest amount of space, but skimping on the floor area in other situations, where the occupant "only" suffers

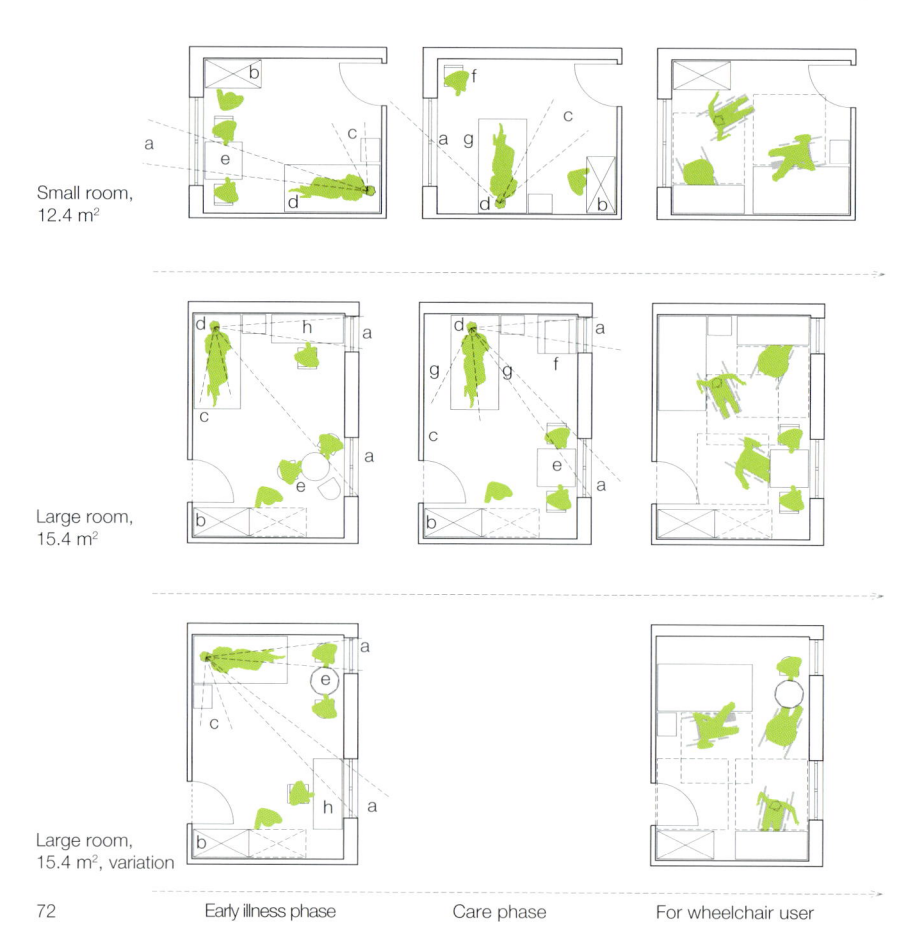

Small room, 12.4 m²

Large room, 15.4 m²

Large room, 15.4 m², variation

72   Early illness phase   Care phase   For wheelchair user

Design
Housing

73

from a motoric impairment, e.g. is reliant on a walking frame, will have a negative effect on the quality of that person's life which is often underestimated. The following movement areas and clearances must be allowed for. A square area measuring 150 × 150 cm enables turning within the room in order to be able to enter and leave the room forwards. With a suitable layout, this movement area can also provide enough room to approach wash-basin and shower forwards, and to transfer to and from bath and toilet. Individual differences may change the need for certain areas or their recommended dimensions considerably, but this aspect is not covered in the standards, which deal with normal situations. For example, there are a number of practical ways of transferring from wheelchair to toilet that do not require any space to the side of the WC, e.g. transferring from the front or slightly diagonally from the front (Fig. 74). However, in most cases transferring from wheelchair to toilet requires a space 90 cm wide and 70 cm deep to one side of the toilet. The wheelchair user reverses into the "parking space" adjacent to the toilet and with the help of one support rail pulls himself/herself forwards and past the side of the wheelchair, in a slightly upright position, onto the toilet seat. To save space and also to allow the toilet to be used by others not reliant on a wheelchair, the support rails can be folded against the wall. This often dictates the side for the transfer procedure, the user cannot choose left or right; the load on the arms in particular therefore cannot take individual requirements into account. Nevertheless, in housing a one-sided approach is normal. Otherwise identical dwellings can be designed with mirrored arrangements so that potential tenants have the choice. For wheelchair users, the clearance to the other side wall or another sanitary appliance on the other side of the toilet should be 35 cm.

The movement area for non-wheelchair users is 120 × 120 cm. Approaching the toilet from the side is also unnecessary. A comparison of bathrooms with a shower for wheelchair users and other users reveals how much extra space is required (Fig. 75).
Standard toilets are between 53 and 58 cm deep, i.e. are not deep enough for the transfer to and from a wheelchair. Such WCs must be 70 cm deep. If a standard toilet is to be used, it must be mounted on a projecting wall, which is no wider than the toilet itself, in front of the actual wall such that the distance from wall to front edge of WC is 70 cm.
Complying with the required clearances and the correct positioning of the sanitary appliances within the room are important aspects, but it is also necessary to provide an adequate number of strong fixing points for support and grab rails. The end of any such rail must be able to carry a point load of at least 1 kN and the fixings must be really secure. Special strengthening measures will be necessary in the case of lightweight internal walls and dry linings. The recommended height for the toilet, 46–48 cm, matches the normal seat height of a wheelchair. This height can be adjusted by fitting an appropriate toilet seat riser. Reaching the toilet paper can be a problem when the toilet is not directly adjacent to a side wall – and not only for disabled persons. One option for the correct positioning of the toilet roll holder is to integrate one into the front end of both support rails. This is also the ideal position for locating a switch that operates an electrical toilet flush.
A floor area of 150 × 150 cm is also necessary for a shower for a wheelchair user, to allow a person to manoeuvre, move and clean themselves properly. Folding grab rails and a tip-up seat, so the user can switch to and from a wheelchair if desired, ease the use of the shower; indeed,

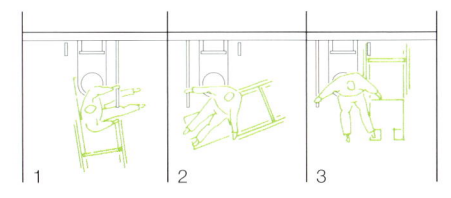

74

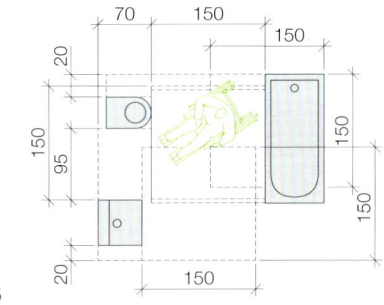

75

Children's centre, Innsbruck (A), 2008, Nickl & Partner
71 Markings on all-glass elements with light/dark contrasts
73 The spandrel panel height of max. 60 cm permits a direct view of the outside world from every patient's bed.
72 Variability in room layouts, housing competition for dementia patients, Munich (D)
    a View out
    b Storage
    c Overview of room
    d Sleeping
    e Communicating
    f Sitting
    g Personal hygiene
    h Writing
74 The figures indicate the three most usual positions for wheelchair/WC transfer.
75 The installation of a shower should still allow for the installation of a bath, with the necessary movement area, at a later date.

Design
Housing

76 a

b

77

78 a

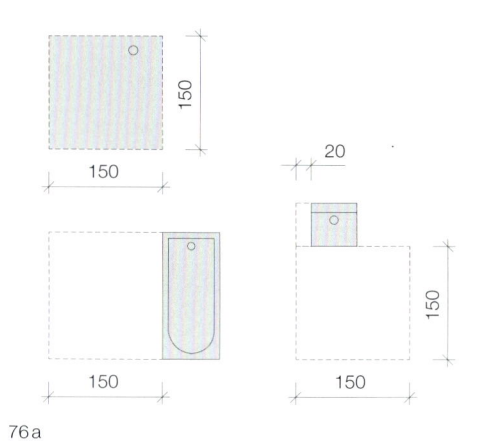

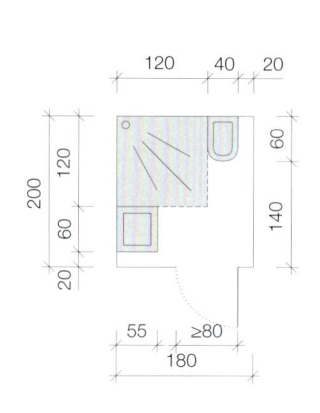

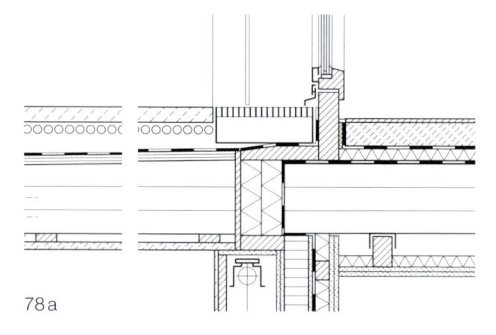

b

79

without such facilities, it may be impossible for a wheelchair user to take a shower. The recommended position for the tap is 85 cm above the floor of the shower, with 50 cm clearance to any corner.

Adequate space must be provided in front of a bath so that an old or disabled person can easily transfer to a bath. An area 150 cm deep alongside the bath is necessary for this. The bath rim should be about 50 cm high, i.e. not much higher than the seat height of a wheelchair. A bath rim as low as possible is recommended when the bath is to be used by older people because this considerably reduces the risk of falling. Steps in front of the bath can help entry and exit, but at the same time could represent an additional risk in themselves. The provision of such steps depends very much on the individual case.

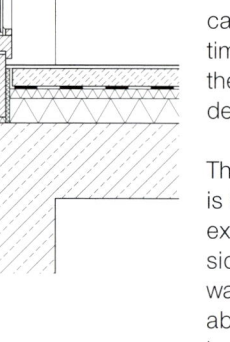

The recommended width for a wash-basin is 60 cm – normally a custom size that, for example, provides wider surfaces at the sides for support. The underside of a wash-basin should not be lower than 67 cm above floor level so that there is enough legroom underneath for independent or assisted cleaning from a seated position. A shallow trap will have to be fitted to achieve this and the minimum depth below the wash-basin will need to be 55 cm so that a seated person can position himself/herself directly adjacent to the rim of the wash-basin.

Severely disabled wheelchair users and the frail elderly cannot usually operate a tilting mirror, so a mirror that begins at the top edge of the wash-basin should be preferred. But in order that a standing person can still use the mirror, it will need to be about 100 cm high (and the same width as the wash-basin itself, i.e. 60 cm). Providing a non-slip floor covering in a wet area is plain common sense. Barrier-free WCs and bathrooms must be provided with windows that can be opened by their users. But a mechanical ventilation system is more hygienic because disabled or older people take longer on average. During cold weather in particular, a window cannot remain open for very long. It is precisely for this reason that an additional sanitary space with WC is advisable in households with several persons.

Balcony, patio, outdoor seating
An outdoor area near to or directly attached to a dwelling is especially important for persons with restricted mobility. Such an area should measure at least 4.5 m² and should include at least one part measuring 150 × 150 cm to allow manoeuvring. Any existing solid balustrade should permit a view through above a level of 60 cm above floor level so that a person in a seated or lying position can still enjoy a view of the outside world.
At the interior-exterior transition – generally at external doors – the understandable desire to omit the threshold does pose technical problems. According to the relevant directives covering sealing against the ingress of moisture at external doors, the distance between the top edge of the moisture barrier and the horizontal building joint should be 15 cm at the door opening. The idea of this is to prevent a build-up of water entering the building. However, the maximum step that can be negotiated safely by a wheelchair user or other mobility impaired person is 2 cm (and less would be even better). One solution to this conflict is to provide a drainage channel directly in front of the door. But even this is difficult to integrate in full compliance with the regulations without providing additional measures, e.g. overhanging eaves in order to reduce the amount of precipitation that could collect. Architects are therefore recommended to advise their clients about this problem.

Design
Possible exceptions

80

Current legislation does not permit full conformity with the regulations (Figs. 78a, b).

Windows
The sizes and clear opening dimensions of the windows required in habitable rooms for lighting, ventilation and a view out are laid down in building regulations. Their vertical and lateral positioning (with respect to internal walls) is not, however, specified.
The barrier-free opening and closing of windows and the opportunity for a view of the outside world result in requirements regarding the positioning and features of windows. The height of handles (and operating devices for high-level lights) above floor level is particularly relevant. Considering a person who has to open/close a window from a wheelchair, then any handle or other device must be located at a height of 85–105 cm above floor level. This leads to technical problems with the tilt function, especially with very tall windows. It is either necessary to abandon the tilt function completely or provide mechanical assistance. If a window extending down to floor level can be fitted instead of the normal situation of spandrel panel plus window above, then this problem is solvable.
The mounting height of windows is also important: occupants must be able to enjoy a view of the outside world from a seated, possibly even a lying position. This means that opaque spandrel panels below windows cannot be any higher than 60 cm above floor level. But such a height does not comply with building legislation, which prescribes a safety barrier 90 cm high, in some situations 100–105 cm. the provision of glazing down to floor level with an external safety barrier is a good solution (Fig. 80).
Being able to approach close enough to doors and windows and their operating handles and locks, is another important criterion; their positioning with respect to lateral obstacles such as walls must be considered. A clearance of 50 cm is necessary in order to guarantee that wheelchair users are catered for (p. 55, Fig. 46).

Possible exceptions
There is no rule without an exception, and despite all the definitions in the Model Building Code (MBO) and the almost identical formations in the building regulations of the federal states, there is an exemption clause concerning building without barriers. The stipulations of MBO cl. 50, paras.1–3 do not apply when (justified) disproportionate additional costs and work would ensue due to the following:
· Unfavourable terrain conditions
· An unfavourable existing building
· The safety of disabled or older persons is impaired

Deviations and exceptions are in principle permissible at the discretion of the particular authority responsible.

References:
[1] Rau, 2008
[2] DIN 32984 "Ground surface indicators in public areas", 2000
[3] GFUV: Workshop on ground surface indicators, 2008

76 a Dimensions of movement areas suitable for wheelchair users adjacent to shower, bath and wash-basin
    b Dimensions of barrier-free sanitary space
77 Floor-level shower
78a, b Stepless access to balcony or patio through the installation of a drainage channel and a timber grating; the inclusion of such a grating mitigates the reduced protection against splashing water.
79 Stepless interior/exterior transition, Solar Decathlon, Darmstadt (D), 2007, Hegger
80 Inward-opening, secure full-height glazing to all rooms, Herderpark, Bad Tölz (D), 2008, Goetz Hootz Castorp

69

Design

T5: Main content and dimensional stipulations of DIN 18024, 18025 and 18040[1,2]
The terms, content and dimensions used correspond to the definitions in the standards. Planning stipulations for wheelchairs are designated with (w). Expressions in italics are those in DIN 18040, which was only available in draft form at the time of going to press.

Movement areas Movement areas may overlap, apart from doors to lift shafts.	• Pedestrian refuge on traffic island or central reservation	W 400 cm × D 250 cm
	• in footways in the vicinity of nurseries, schools, leisure or care facilities • in pedestrian crossings and other crossing places	W 300 cm
	• Pedestrian refuge in pedestrian crossings and other crossing places	W 300 cm × D 200 cm
	• in footways alongside local distributor roads	W 200 cm
	• Turning option (except small rooms that wheelchair users can use without limitations either forwards or backwards [W]) • Rest area • at the start and end of a ramp • in front of house and building entrances • in front of controls • in front of service counters, aisles, checkouts and security barriers • at the start and end of an escalator • in front of lift doors • adjacent to outdoor seating • for showering (W) • in front of WC (W) • in front of wash-basin (W) • in front of dustbins or rubbish chutes (W)	W 150 cm × D 150 cm
	• in footways and main paths • at the start and end of a flight of stairs (last step top or bottom may not be included in movement area) • between walls outside dwellings • in corridors	W 150 cm
	• along the full length of a vehicle • in front of therapy apparatus and facilities • along the full length of a bed for a wheelchair user (W) • in front of cupboards (W) • in front of kitchen appliances/worktops/cupboards (W) • along the full length of a bath (W) • in front of a wheelchair parking space (W)	D 150 cm
	• in front of sanitary appliances • in a floor-level shower area	W 120 cm × D 120 cm
	• on main paths (but depends on particular situation) • clear width between the kerbs of a ramp • alongside facilities used by wheelchair users in buildings accessible to the public • on paths within a residential complex • between walls within a dwelling (also [W]) • in front of kitchen appliances/worktops/cupboards • along the full length of a bed that must be accessible from three sides • alongside furniture that wheelchair users must approach sideways (W) • along the full length of a bed for a non-wheelchair user (W) • adjacent to controls (W)	W 120 cm
	• Aisles, e.g. between checkouts or security barriers • in secondary paths	W 90 cm
	• in front of furniture	D 90 cm
	• alongside stops for local public transport	D 250 cm
Passing places	• on main and secondary paths and footways within the range of vision, max. spacing 18 m	W 200 cm × D 250cm W 180 cm × D 180 cm
	• in corridors every max. 15 m	W 180 cm × D 180 cm
Pedestrian circulation areas	• Safety verge between footway and carriageway of main road • Delineator strip between cycle track and footway	W 75 cm W 50 cm
	• Difference in height between carriageway and footway in access and local distributor roads • Kerb must be lowered at access points and pedestrian crossing points	> 3 cm < 3 cm
	• Longitudinal gradient of footways without rest areas • Longitudinal gradient of footways with rest areas every 10 m • *Longitudinal gradient of footways up to 10 m long* • Transverse fall (at access points max. 6 %)	3 % 3–6 % 4 % 2 %/2.5 %
Main footways	• Clear opening (a width of 120 cm may be adequate over a length of 200 cm in certain situations)	W 150 cm × H 230 cm
	• Longitudinal gradient • Longitudinal gradient in exceptional situations where there is a rest area every 10 m • Transverse fall	4 % 4–6 % 2 %/2.5 %
	• Spacing of seating	100 m
Secondary footways to barrier-free playing and recreational apparatus plus adventure areas	• Clear opening	W 90 cm × H 230 cm
	• Longitudinal gradient • Longitudinal gradient in exceptional situations where there is a rest area every 10 m • Transverse fall	6 % 4–6 % 2 %
Securing of building sites	• Barriers 10 cm high • Additional tactile bar 10 cm high	top 100 cm above FFL top 25 cm above FFL
	• Clear opening	W 120 cm × H 230 cm
Local public transport stops/stations	• Difference in height and clearance between interior of vehicle and adjacent access level	< 3 cm

[1] Part 1 has been approved and is due to be published in the late summer of 2010. [2] Part 2 is due to be approved soon and should also be published in the late summer of 2010.

Design

Car parking spaces	• Parking parallel to road • *Parking space for wheelchair user (parking at 90° to road)*	L 750 cm × W 250 cm *W 350 cm × D 500 cm*
	• In public places, parks, gardens and playgrounds, 3 % of the parking spaces, but at least 1 space, must be suitable for wheelchair users in accordance with DIN 18025-1. • In parking areas for buildings accessible to the public, 1 % of the parking spaces, but at least 2 spaces, must be suitable for wheelchair users and located near the main entrance. • 1 parking space for a minibus must be included near the main entrance of a building accessible to the public.	L 750 cm × W 350 cm, H 250 cm
	• 1 covered parking space or garage must be provided for every dwelling intended for a wheelchair user (W).	
Floor finishes outdoors	• Readily accessible with a wheelchair without undue vibration	
Floor finishes indoors	• Non-slip, suitable for wheelchairs and permanent (according to ZH 1/571, now BGR 181), with no risk of a build-up of static electricity *(high-contrast design)*	*non-slip,* *min. class R9 to BGR 181*
Doors	• Clear opening	W 180 cm × H 210 cm *H 205 cm*
	• Movement area in opening direction of a side-hung door • Movement area facing opening direction • Movement area in front of sliding door, both sides • Movement area in front of lift door	W 150 cm × D 150 cm W 150 cm × D 120 cm W 190 cm × D 120 cm W 150 cm × D 150 cm
	• Door threshold	< 2 cm
	• Doors to WC, shower and changing cubicles may not open inwards	

	• High-contrast design (wall/frame/leaf ...)
	All-glass doors must be provided with safety markings that
	  - extend over the full width,
	  - are visually contrasting, and
	  - have light and dark components (alternating contrast)
		H 40–70 cm and
		120–160 cm above FFL

	• *Door viewer in dwellings intended for wheelchair users (W)*	*H 120 cm above FFL*

	• Easy-to-use door opening/operating hardware by way of...
	  - curved or U-shaped handle
	  - vertical bar for manually operated sliding door	

Stairs Spiral or curved stairs are not permitted for compulsory staircases. Projecting nosings (i.e. set-back risers) are not permitted.	• *Handrails on both sides* • Outer handrail continues beyond start and end of stairs • *Handrail ends that project into open space must be finished off with a bend downwards or to the side.* • *Clear distance between handrail and wall (note: DIN 18065 specifies 4 cm)*	30–45 mm dia, top 85 cm/*top 85–90 cm* 30 cm 5 cm
	• Kerb at side of open steps	H 2 cm
	• Unobstructed height clearance beneath stairs	H 230 cm/*H 220 cm*
	• *Straight flight(s) essential (exception: stairwell diameter > 200 cm)*	
	• must include risers • *Angled risers (to increase tread depth) are permitted*	*< 2 cm*

	• First and last step visually contrasting over entire width of stair
	• *Marking elements on steps, for example:*
	  - continuous strips on tread
	  - directly on nosings
	  - on ends of risers
	  - clear contrast between tread and riser
	• *Stairs that begin in an open space require tactile information surfaces directly in front of the bottom-most step and directly after the topmost step*
		W 5–8 cm
		
		W 4–5 cm
		W > 1 cm
		
		D > 60 cm

Escalators	• Speed	< 0.5 m/s
	• Inclination • Gradient of passenger conveyor	< 30° < 7°
	• *Every step must be marked* • *Entry and exit points marked with strips*	*W 8 cm*
Ramps	• Longitudinal gradient, without transverse fall	< 6 %
	• Length between intermediate landings (150 cm long)	600 cm
	• Kerbs on both sides of ramp and intermediate landings	H 10 cm
	• Clear width between kerbs	120 cm
	• Handrails on both sides	30–45 mm dia, top 85 cm
	• Handrails continue horizontally beyond start and end of ramp and intermediate landings	30 cm
Lifts	• Clear dimensions of lift car	W 110 cm × D 140 cm
	• Lift shaft door	W 90 cm
	• Distance between lift shaft door and descending stair • Movement area in front of lift shaft door	min. 300 cm 150 cm × 150 cm
	• Fitted with mirror, grab rail and horizontal controls suitable for wheelchair users • with acoustic signals if required	
Controls	• Fixing height	centre-line 85 cm above FFL
	• *Several controls, one above the other* • Lateral clearance between controls and wall • *Legroom for frontal operation*	*centre-line 85–105 cm* > 50 cm *55 cm*

Design

(Tab. T5 contd: Main content and dimensional stipulations of DIN 18024, 18025 and 18040)

Controls (contd.)	• Readily identifiable through tactile features and visually contrasting design • *High visual contrast and tactile features (principle of two senses)* • *Functions must be identifiable*	
	• not recessed, no sharp edges	
	• Exclusive use of touch controls is not permitted • *Exclusive use of touch controls, touchscreens or contactless controls is not permitted* • *Acknowledgement of activation of each function*	
	• Clearance between switch for operating non-manual door and door opening space	> 250 cm, opposite side > 150 cm
	• Safeguards to prevent injuries if there is a risk of catching fingers or clothing; rounded edges and rubber guards to reduce risk of injury (W)	
	• Emergency call switch also operable from floor level (pull cord)	
	• Water temperature without additional scalding protection	< 45 °C
	• Radiator valve	H 45–85 cm above FFL
	• Name-plates at building entrances should be provided with tactile, embossed labelling.	

Fitting-out, orientation, signs and lighting	• Legibility of equipment and machines by way of…
	  - max. 10 cm high set-back plinth,
	  - 3 cm high "silhouette" or
	  - tactile bar 15 cm high
	  matching the size of the equipment/machine
		H 3 cm
		
		top 25 cm above FFL

	• Illumination of circulation areas and stairs without glare and without shadows (higher illuminance than given in DIN 5035-2)	
	• Escape route signs supplemented by lighting strips, lighting to indicate direction and audible signals	
	• Tactile information at start and end of stair handrail	
	• Lifts with more than two stops require audible announcements	
Sanitary facilities	• At least 1 publicly accessible sanitary facility suitable for wheelchair users (to DIN 18024-2) required in parks and leisure facilities	
WC	• Movement area to left and right of WC • Movement area in front of WC • WC height (incl. seat) • User must be able to lean back (WC seat cover unsuitable for this)	> W 95 cm × D 70 cm *W 90 cm × D 70 cm* W 150 cm × D 150 cm 48 cm above FFL *46–48 cm above FFL* 55 cm behind front edge of WC
Support and grab rails	• to left and right of WC pan, fold-up *(in user-defined stages)*, lockable in horizontal and vertical positions, projecting 15 cm beyond front edge of WC pan	spacing 70 cm, top edge 85 cm above FFL, 28 cm above seat *spacing 65–70 cm*
Toilet flushing	• in support rail on both sides	
Toilet roll holder	• 1 holder at front end of each support rail	
Wash-basin	• Level of rim above FFL • Room for knees/legs underneath, with shallow trap or trap concealed in wall • Knee room • Movement area in front of wash-basin • Single-lever or contactless tap	< 80 cm *clear space underneath over width of 90 cm* D 30 cm × H > 67 cm W 150 cm × D 150 cm
Mirror	• above wash-basin, also usable from the seated position	H > 100 cm
Soap dispenser	• Dispensing point	85–105 cm above FFL
Hand drier	• approachable with wheelchair • Height of air outlet • Movement area	85 cm above FFL W 150 cm × D 150 cm
Waste receptacle	• Height of sealed, odour-tight waste receptacle with self-closing cover	85 cm above FFL
Miscellaneous	• Tap/valve with hose and floor outlet • Emergency call button that guarantees immediate help; also operable from floor level (pull cord) • Clothes hooks • Sanitary rooms, e.g. in roadside service areas or sports centres, to be fitted with folding adult changing table. • *In sanitary rooms ventilated exclusively via windows it must be ensured that the windows can be opened/closed from a wheelchair.*	H 85 cm and 150 cm above FFL table 200 × 90 cm, H 50 cm above FFL *table 180 × 90 cm, H 46–48 cm, movement area D 150 cm handle height 85–105 cm*
Toilet or shower cubicles	• Shower without thresholds (W)	W 150 cm × D 150 cm
	• with tip-up seat for showering	W 40 cm × D 45 cm top 48 cm above FFL *top 46–48 cm above FFL*
	• with fold-up support rails both sides	28 cm above seat
	• Soap dish	top 85 cm
	• Single-lever tap	85 cm
Changing cubicles	• At least 1 changing room for wheelchair users in workplaces, sports centres, swimming pools and therapy establishments	
Wheelchair parking space	• Parking space preferably near the entrance (1 space with battery-charging point to DIN VDE 0510 part 3 for every wheelchair user [W])	W 190 cm × D 150 cm *W 180 cm × D 150 cm*
	• Movement area in front of wheelchair parking space	W 190 cm × D 150 cm *W 180 cm × D 150 cm*

Design

Places of assembly, sports facilities and restaurants	• Spaces for wheelchair users	W 95 cm × D 150 cm
		W 90 cm × D 130 cm, adjacent movement area 150 cm deep
	• *Spaces for wheelchair users with sideways approach*	*D 150 cm × W 90 cm, movement area to side of this 90 cm wide*
	• 1 % of spaces, but min. 2 spaces for wheelchair users; seats for accompanying persons adjacent to wheelchair spaces	
	• *Seats with greater legroom should be included for larger persons and those with mobility impairments.*	
	• *Permanent tables (lecture theatres) require writing surfaces for wheelchair users; if electroacoustic PA systems are in use, then a PA system for the hearing impaired, which serves the entire audience area, must be provided.*	
Places of accommodation	• 1 %, but min. 1 room suitable for wheelchair users to DIN 18025-1, must be included.	
	• All devices (incl. curtains, door locks, etc.) should be designed for remote operation if necessary.	
Counters, ticket windows and sales tables	• Suitable for wheelchair users	85 cm above FFL
		80 cm above FFL
	• with clearance for knees/legs at one point at least	D 30 cm × H > 67 cm
		room for knees/legs over width of min. 90 cm with D > 55 cm
	• *Size of movement area can be reduced if counter has room for knees/legs over a width of 150 cm.*	W 120 cm
	• *Service counter with closed glazing and intercom, also ticket windows and checkouts, in loud environments must also be fitted with an audio induction loop for handling confidential matters.*	
Special requirements relevant to housing for wheelchair users according to DIN 18025-1 (W)	• All rooms belonging to the dwelling and the common facilities of the residential complex must be accessible without having to negotiate steps, if necessary via lifts or ramps.	
	• All dwellings not suitable for wheelchair users must be made accessible by retrofitting a lift or ramp.	
	• Door threshold	< 2 cm
Kitchen	• Cooker, worktop and sink must be fully accessible with room for knees/legs underneath and should be positioned across a corner.	
Bathroom	• Shower suitable for wheelchair user	
	• It must be possible to retrofit a bath with transfer seat.	
	• Wash-basin with room for knees/legs underneath	48 cm above FFL
	• WC height (incl. seat)	*46–48 cm above FFL*
	• Mechanical ventilation	
	• An additional sanitary room (with WC and wash-basin at least) must be provided in dwellings with more than three occupants.	
Additional living space	• Additional living space may need to be provided for wheelchair users. The appropriate living area increases according to cl. 39, para. 2, 2nd Housebuilding Act and cl. 5, para. 2, Publicly Assisted Dwellings Act.	in normal case by min. 15 m²
Outdoor seating	• Every dwelling shall be allocated an outdoor seating area.	> 4.5 m²
	• Movement area	W 150 cm × D 150 cm
Additional requirements for barrier-free housing according to DIN 18025-2	• The building entrance and one dwelling level shall be accessible without having to negotiate steps.	
	• All rooms belonging to the dwelling and the common facilities of the residential complex must be accessible without having to negotiate steps, at least by the retrofitting of a lift or ramp.	
Bathroom	• Floor-level shower	
	• Room for knees/legs below wash-basin	
Additional living space	• Additional space may need to be provided, e.g. for small persons, the blind and the visually impaired. The appropriate living area increases according to cl. 39, para. 2, 2nd Housebuilding Act and cl. 5, para. 2, Publicly Assisted Dwellings Act.	in normal case by min. 15 m²
Outdoor seating	• Every dwelling shall be allocated an outdoor seating area.	> 4.5 m²
	• Movement area	W 150 cm × D 150 cm
Spandrel panels	• should provide a view through in at least one habitable room in the dwelling and from outdoor seating	from H 60 cm above FFL
Room temperature	• Heating of dwellings and common facilities must be adaptable to suit individual requirements throughout the year.	
	• Water temperature	< 45 °C
Warning, orientation, providing information	• *Important information must be provided according to the principle of two senses.*	
	• *Information can be provided by visual, audible or tactile means.*	
Factors influencing visual information	• *Luminance contrast (light/dark)*	
	• *Size of visual object*	
	• *Form*	
	• *Spatial arrangement, positioning of visual object*	
	• *Viewing distance*	
	• *Colour contrast is not a substitute for luminance contrast.*	
Factors influencing acoustic information	• *Avoid disturbing noises in rooms*	
	• *Protect against external sources of noise*	
	• *Avoid long reverberation times*	
Factors influencing tactile information	• *Tactile information in the form of raised lettering but also in Braille*	
	• *Access to separate-sex facilities, e.g. WC, shower, changing room, to be marked appropriately, also room doors*	
Alerting and evacuating	• Fire safety concept must take into account the needs of persons with motoric and sensorial impairments, for example:	
	- by providing safe refuge areas for those persons not able to escape/rescue themselves	
	- additional visual alarms in addition to acoustic alarms and warning signals	
	- operational precautions.	

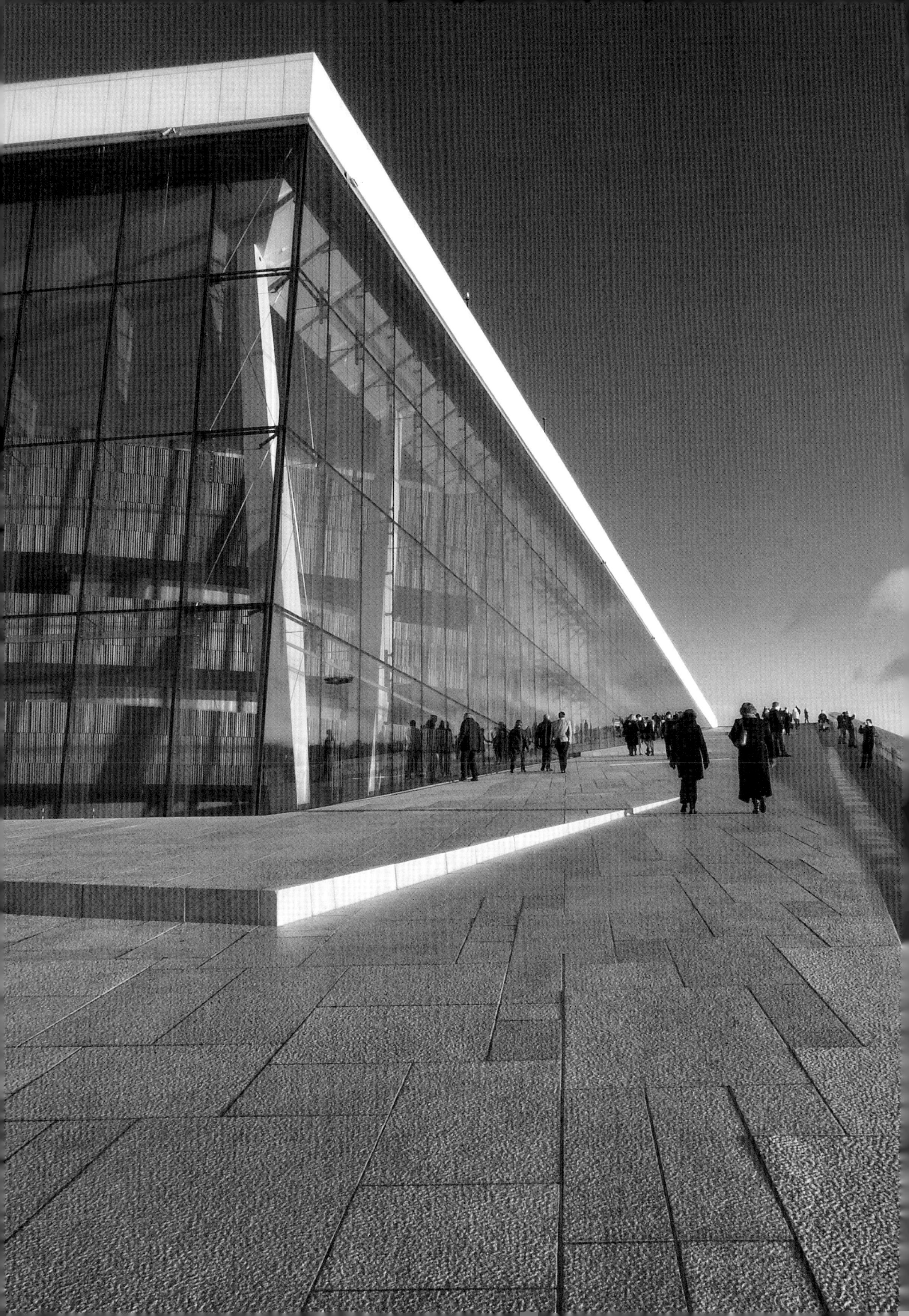

Typology

"The third teacher"
Learning calls for inquisitiveness. Growing up has much to do with exploring and conquering the world, questioning and disobeying rules, asserting oneself in the world of grown-ups. Besides the pure dissemination of information, "learning how to learn" is one of the most significant tasks of any school.
According to a Scandinavian proverb, the space is the third teacher. (The first teacher is the other children, the second teacher is the tutor.)

It is regularly attested that the German education system does not really achieve integrative environments. In no other EU Member State is the course of a child's education, and hence frequently its course in later life too, so unequivocally dependent on parental status.
In order to be able to demonstrate supportive options for this system, which is anything but barrier-free, it is worth asking the question as to what we regard as "building for children". The following are some of the influencing factors that need to be considered:
- Supporting natural inquisitiveness
- Encouraging inquisitiveness in an environment that enables comparatively safe actions (safety)
- Promoting integration
- Promoting a sense of identity
- Demonstrating respect for the needs of the users
- Providing orientation, but without constraints
- Responding to the size of children at appropriate places

The children who are good at mathematics are not those who complete extra exercises, but rather those who can also balance well on a kerb. But instead of allowing children to climb trees, they are required to absorb ever more mathematical content. [1]

The potential inherent in schools is realised totally differently in the Baupiloten projects. In a modelmaking workshop at the Carl

Carl Bolle Primary School, Berlin (D), 2008, Baupiloten
1   In the "Spy Wall" the children can climb, hide and still maintain a view over the entire corridor.
2   Interior layout of the school with its imaginary worlds
   a Periscope
   b Afterglow
   c Codes
   d Complementary colours
   e Flash-puffs
   f Reading hatches
   g Spy cell
   h Criss-crosser
   i Rainbow spectrum
   j Light conductor
   k Sensitive listening
   l Two-sidedness
   m Twinkling gate

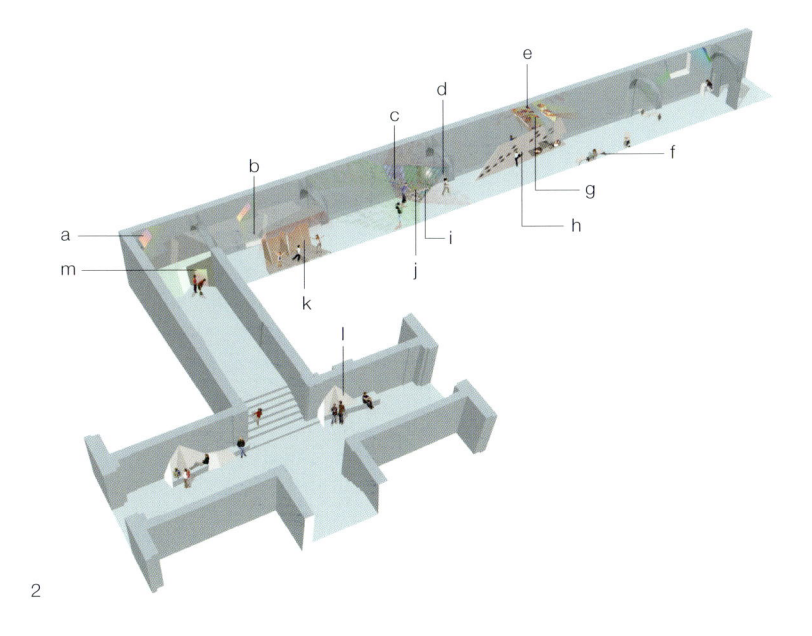

Typology
"The third teacher"

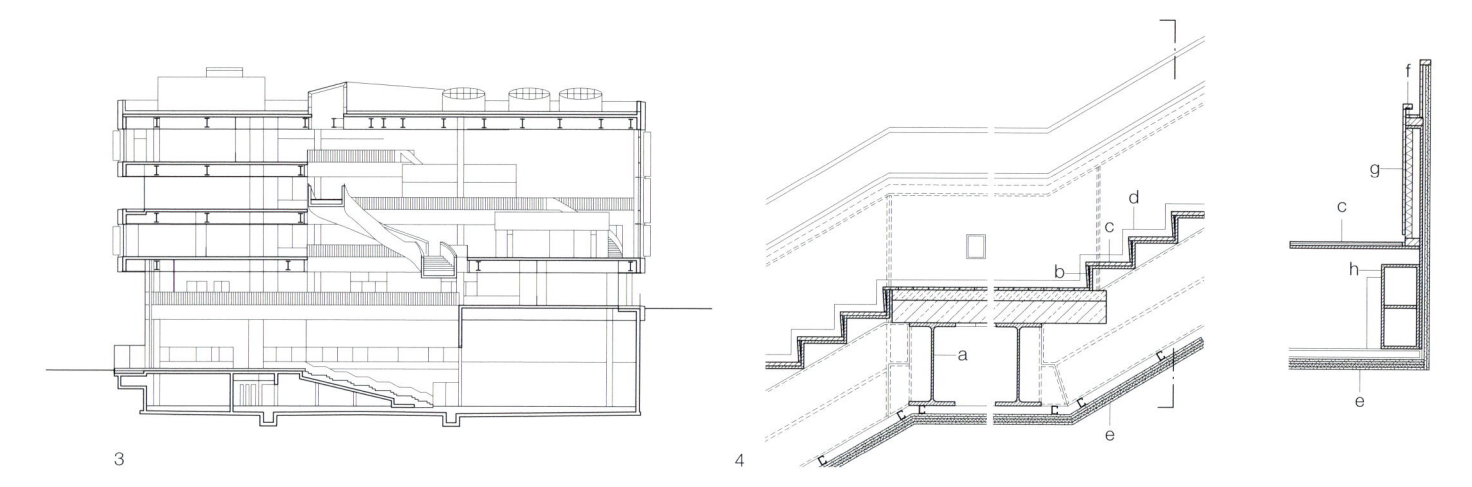

Bolle Primary School in Berlin, the Baupiloten team, together with the pupils, developed imaginary worlds in non-teaching and circulation areas only. The names given to these areas ("Summer Labyrinth", "Climbing Forest", "Secret Leisure Garden" and "Snow World") provide clues to their usage, conception and atmosphere (p. 75, Figs. 1 and 2). Baupiloten is a fluctuating group of architecture students, guided by professional architects, that independently develops, designs and arranges the construction of building measures during their studies. Their involvement results in places that encourage the children to become active in a playful way, but which are also places where they can escape from the daily turmoil.

The differentiated interior design of one connecting corridor, the "leisure area" of the Carl Bolle Primary School, stimulates the children's perceptions. The specially designed spaces encourage independent activity, orientation, communication, coexistence, physical experiences and a feeling for aesthetics. The space has become an area for research and experimentation in which the children can develop a picture of themselves, the other children and the world (p. 75, Fig. 2).
The goal is to integrate the experimental corridor zones into the curriculum, or rather to be able to reflect on the experiences gained there in the lessons, in the form of "discovery learning". The outcome is not only imaginary worlds and spaces, but also three-dimensional environments – in what are ostensibly non-teaching areas – that permit discoveries, relaxation, play and a wealth of joint learning.

This notion is taken a step further in the design of the Ørestadt Grammar School in Copenhagen. Circulation and teaching zones have been merged within the building to form a single, multi-purpose communication area. Besides a few traditional, enclosed group rooms, the three-storey, open hall includes teaching and learning "islands" that result in a totally new form of community within the educational environment, without any defined outward constructional framework (Figs. 3–7).

Following the Danish grammar school reform, pupils between the ages of 16 and 19 are no longer divided into separate academic years, but rather rely on focused, self-organised learning in teams and groups. The visible physical translation of this in Ørestadt is revealed as an ideal form of joint study.
Physical barriers seem no longer to exist. However, unfamiliar acoustic situations can occur. In order to minimise disturbances, the community of pupils must not only discover itself as such, but also develop responsible forms of co-existence.

The above examples show spatial environments that help children to learn from and with each other. The open spaces create integration options that are not available in many educational establishments, at least not in such an obvious way. They promote – almost casually – social and integrative skills to the same extent as the joint development of different learning strategies.
The adults must show a large degree of respect towards the pupils, and the

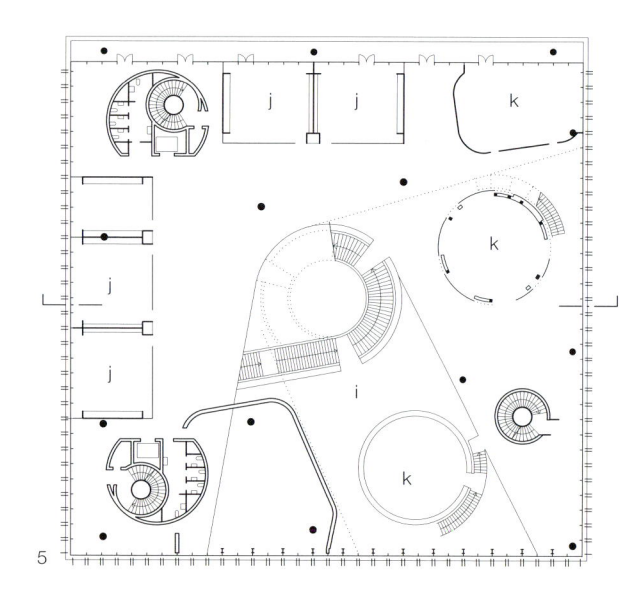

Typology
"The third teacher"

pupils must show a similar respect towards one another. It is against this background that we must assess the new dining hall at the Luisen Grammar School in Munich, which illustrates how respect can contribute to an environment for schoolchildren (Fig. 8).

The regular, comparative studies of the Organisation for Economic Cooperation & Development (OECD) at European schools have instigated an ongoing debate – in Germany too – about school forms, teaching methods and learning content. In many schools this has already made itself felt in the change to all-day schooling. If we can avoid focusing purely on the functional necessities, e.g. the provision of meals at lunchtime, and instead develop schools into places of identification, networking and responsibility, then this process will bring us close to the good Scandinavian models.

The internal spaces at the Luisen Grammar School in Munich enable us to draw parallels with the Ørestadt Grammar School in Copenhagen. Here, we see the creation of very flexible spaces with furnishings and fittings that permit great degrees of freedom. These are not just spaces for lunchtime relaxation and supervised activities. The form and design encourage self-organised learning groups and teams as in the Danish example. The configuration of the internal spaces, furnishings and fittings is remarkable. All furniture includes legroom for wheelchair users and all the seating is easily approached by wheelchair users. Consequently, any potential obstacles to barrier-free design are avoided from the outset.
Materials, forms and colours are clearly contrasting and hence distinguishable. It is immediately obvious to users that these internal spaces are specially designed, with specially produced fittings, not standard elements. The quality of layout and materials create an atmosphere more like a lounge than a school dining hall.
When the design of the building itself allows a school to become "our school", then the pupils can identify with it. This form of respect for the users is something that Maria Montessori recognised and implemented with success (see "Historical review", p. 12).

Ørestadt Grammar School, Copenhagen (DK), 2008, 3XN Architekten
3 Section, scale 1:750
4 Detail section through central, barrier-free stairs, 153 steps each 168 × 280 mm on centre-line, 2550 mm wide, scale 1:50
   a HEB 550 steel section with intumescent paint finish
   b Riser, 15–22 mm ash
   c Tread, 22 mm ash, 2 mm "Korkment" (cork on jute backing), 6 mm steel plate
   d Stainless steel section, 2× 80 mm, brushed finish
   e Acoustic plaster on 25 mm background, 2 No. 13 mm plasterboard
   f Handrail, 34 × 60 mm ash
   g Wood-based board product, 10.5 mm ash, 45 mm mineral wool, 55 mm air space, 10 mm steel plate, 20 mm steel trapezoidal profile plate, 2 No. 13 mm plasterboard
   h Stringer, 560 × 250 × 10 mm steel plate
5 Plan of 2nd floor, scale 1:750
   i Atrium
   j Group room
   k Learning island
6, 7 Interior with teaching and learning islands

8 Dining hall, Luisen Grammar School, Munich (D), 2008, Bodensteiner Fest Architekten

Typology
"The third teacher"

9

Buildings and facilities that are to be used principally by children call for an appropriate response to the design and construction challenge of "designing for children". Even though being involved in and learning about the world of grown-ups is a worthwhile aim, it can still be helpful to take into account the average size of the children in the planning work (Tab. T2).
But the needs of children – for safety reasons alone – are important when planning *all* facilities, not just those intended specifically for children. Designs should also make sure that accessability is addressed and that lift controls, alarm buttons, doorbells, keyholes, letter-boxes, etc. are placed at heights that allow them to be used by children without any assistance. Tab. T1 lists the mounting heights of sanitary appliances to suit different ages.

Special schools or facilities such as the Conductive Education Centre in Munich (Figs. 9 and 10) provide care and support for children so that they can achieve a worthwhile school-leaving qualification and have the best possible chance of leading an independent life. The centre in Munich employs education and therapy methods for children with cerebral motor disorders according to the teachings of Prof. Dr. András Petö. Here, therapy and pedagogy are regarded as one and are applied in a holistic fashion. The objective is to achieve minimal dependence on aids (wheelchairs etc.) or outside assistance in everyday activities, in the family and at work. The methods are based on the principle that motor disorders are principally learning disorders, which in addition to causing motoric problems hamper personality development. The centre supports children with physical disabilities in their development of motor, language, mental and social skills. All activities are carried out systematically on an individual basis, and in small groups, and are integrated into everyday activities.
The group of buildings completed in 2005 consists of an educational therapy day centre, primary and secondary schools, a crèche, a kindergarten with school preparation centre, a residential school and a swimming pool. In order to create identifiable orientation options, the buildings, built on a common podium that links the ensemble, are grouped and coloured according to their functions. The arrangement of the basic functions in both the entrance zone (Fig. 9) and the finely detailed interior exhibit great sensitivity to and knowledge of the needs of the centre's users. For example, the design for the areas directly in front of the classrooms exceeds the standard requirements with respect to heights, moving areas, legroom under furniture, etc. Additional glazing in each door leaf not only admits daylight into the corridors, but also enables wheelchair users in particular to see obstacles when opening and closing a door themselves (Fig. 10).

For many severely disabled children, their parents' home or their care facility is the key, defining everyday environment. Organisational, constructional and usage-specific responses and support in the interior layout are necessary in order to guarantee identification, orientation and help towards independence in a school environment. Guidance systems or an inherently memorable design without additive elements helps children to handle their environment. One example of a self-explanatory orientation system, in addition to those in use at the Conductive Education Centre, is that at the children's day centre in Großhadern, a suburb of Munich. Here, the entire floor of the entrance hall matches the playing board of a "Ludo" game (Fig. 11). Each

Typology
"The third teacher"

T1: Mounting heights of sanitary appliances to suit different age groups [3]

Sanitary appliance	min. width of movement area	min. depth of movement area	Mounting height above FFL	Children 3–6	Children 7–11	Children 11–15
Single wash-basin	90	55	85	55–65	65–75	75–85
Handrinse basin	70	45	85	55–65	65–75	75–85
WC pan, wall-mounted cistern	80	60	42[2]	35[2]	35[2]	42[2]
WC pan, built-in cistern	80	60	42[2]	35[2]	35[2]	42[2]
Urinal	60	60	65		50	57
Shower tray	80/70[1]	75				
Baby bath	90	75	85–90			
Washing machine/ tumble drier	90	90				
Classroom sink	80	55	85	55–65	65–75	75–85
Workshop sink	90/120[3]	120	85		65–75	75–85
Sink (single/ double)	90/120[3]	120	85–92			
Bucket/cleaner's sink	80	55	65			
Slop hopper	60	55	65			
Bedpan washer	80	120	65			
Wash-basin for wheelchair users	150 × 150	150 × 150	80			
WC pan for wheelchair users	150 × 150	150 × 150	46[1]			

A distance of 75 cm must be provided between sanitary objects on opposite walls, walls themselves and utility spaces.

[1] with corner entry
[2] top of ceramic rim for wall-mounted appliance
[3] with walls on both sides

T2: Anthropometric data (in cm) of children aged 1–14 [2]

Age	1	2	3	4	5	6	7	8	9	10	11	12	13	14
a Stature	75	85	94	102	109	115	122	128	133	138	143	148	153	158
b Eye height	64	74	83	91	96	103	108	111	115	119	124	128	133	137
c Shoulder height	54	63	72	79	85	90	95	98	102	106	110	114	118	122
d Sideways reach	65	77	89	97	104	110	116	122	127	133	138	143	148	153
e Upward reach	90	100	112	121	129	136	142	147	153	159	165	171	177	183
f Downward reach	28	32	35	39	42	45	48	50	52	54	56	58	60	62
g Forward reach	30	36	42	48	52	57	61	63	65	68	71	73	75	77
h Height of handrail	38	40	42	45	49	55	57	–	–	–	–	–	–	–
i Height of table surface	40	42	43	46	48	51	53	55	57	59	61	64	66	68
k Height of seat surface	19	22	25	28	30	32	34	35	36	38	39	41	42	44

– No information available for this age

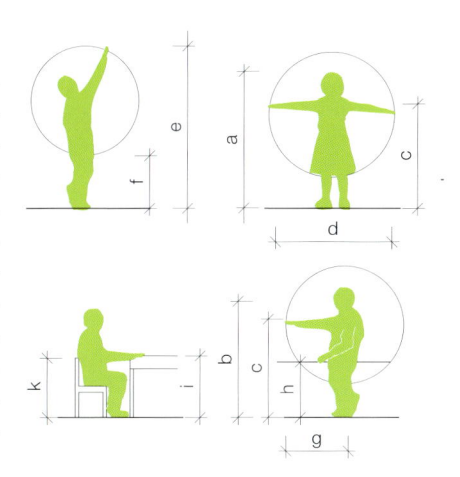

colour stands for one group and hence for that group's room. Neutral colours link the rooms together or to common amenities. So orientation options are supported in a playful manner while stimulating the imagination with the coloured zones which are obviously ideal for a variety of games.

The importance of the "third teacher" should not be underestimated, since its design and construction helps determine the behaviour of the "first and second teachers".

Conductive Education Centre, Munich-Oberföhring (D), 2005, Reichert Pranschke Maluche
9 Lift and stairs enable equal access for all
10 Classroom entrance with door glazing and adjacent cloakroom

11 "Ludo" game as basis for guiding users, children's day centre, Munich-Großhadern (D), 2008, Holzfurtner Bahner Architekten

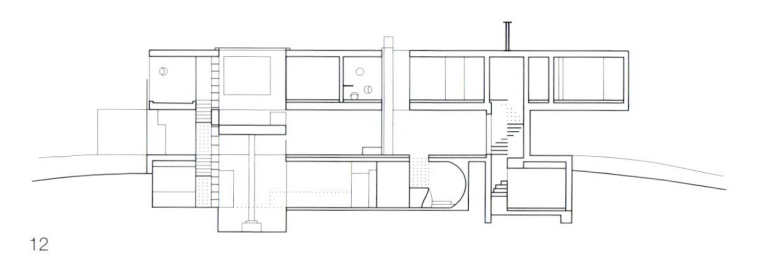

Typology
Housing

12

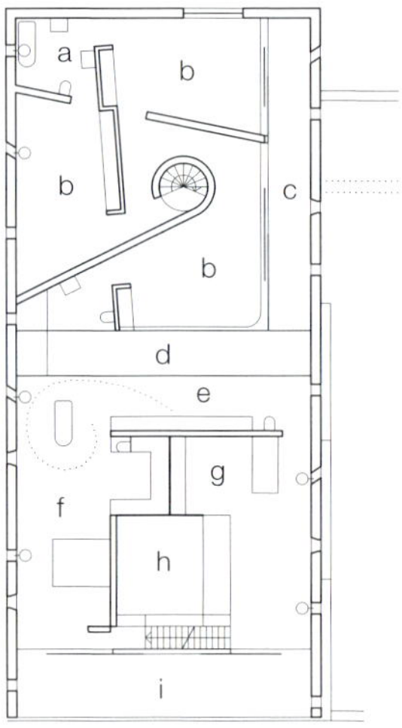

Housing
The housing market in Germany appears to be very standardised and less flexible when it comes to the needs of changing individual requirements. The number of single-person households in different age groups is increasing at a rapid pace. Fluctuations within family structures are the norm. The number of job-related relocations within a person's working life has risen noticeably. These factors combined with demographic developments all have an effect on the housing market. It would be helpful to have a range of products on offer that can be adapted to the diverse needs of different people.
In this respect, the following construction measures should be taken into account when planning new-build and refurbishment projects:
- Barrier-free access
- Movement and manoeuvring areas
- Barrier-free design of handles, switches, etc.
- Barrier-free sanitary spaces
- Barrier-free external amenities

The examples below show how planners can proceed for different typologies.

Detached houses
Providing spaces for individual lifestyles as embodied in the "Maison à Bordeaux" by Rem Koolhaas (1998) will remain a rarity (Figs. 12–15). Nevertheless, this building, which attracted attention on an international scale, does show that the needs of a wheelchair user can be implemented in a very inspirational way. Horizontally, the building is organised on three levels, and vertically in two different access forms. On the lowest level, built into the slope, the private rooms of the users are cut out of the ground like caves, on the topmost floor there are two self-contained areas: one for the children, one for the parents. The almost fully glazed ground floor between these two forms the family's main living area. But the "heart of the house" is a 3.0 × 3.5 m elevating platform, a mobile room with desk that enables the wheelchair-bound client – but not only he – to reach all the floors (Fig. 15).

But design briefs such as this do not necessarily need the input of an architect with an international reputation. Architect Florian Höfer's specification for his first house was very similar. His low-

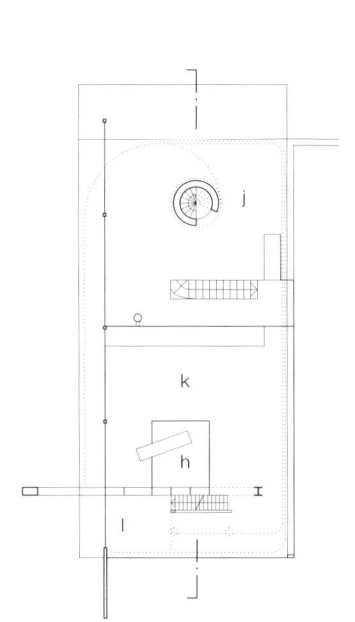

13

14

15

80

Typology
Housing

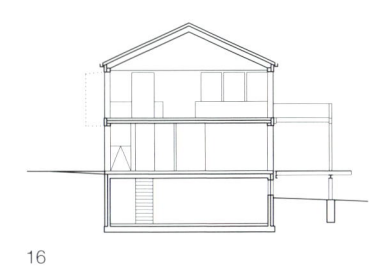

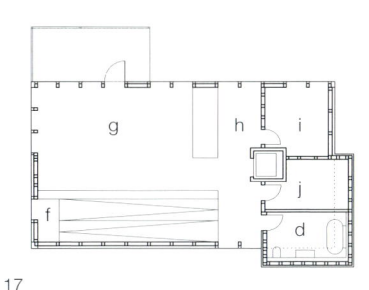

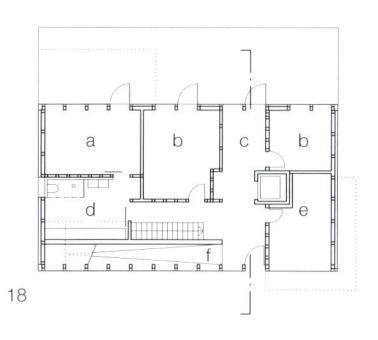

16

17

18

energy house in Gstadt (2004) proves just how natural and unpretentious, but nonetheless memorable, a house designed for a wheelchair user can be (Figs. 16–20). A family's desire to be able to use the area within its own four walls as independently as possible – with three children and a father in a wheelchair – resulted in a concept that turns the standard floor plan typology upside down. The private rooms, bedrooms and bathroom are on the ground floor, the general living area plus study and another bathroom are on the upper floor.
The only permanent fixtures within the living area are a low wall alongside the ramp and the kitchen units. The wall contains shelves, but is also used as casual seating. This fixed element houses light switches, power sockets, etc., too, at a height where they can be used by all occupants. The fixed kitchen units include legroom underneath.
A ramp joins the two levels, and its gradient has a major impact on the overall length of the house. The 10 % gradient was chosen by the client based on his individual capabilities and requirements, but it is also the sign of an inherent,

conceptual intention. The house and its envelope required an open design based on maximum flexibility. With future needs in mind, a lift was also included because the family is relatively certain that this environment, designed around its needs, is the house in which they will grow old.

It would seem that in the meantime building without barriers has become a natural factor in buildings designed for sustainable and energy-efficient usage. For example, in the centre of Memmingen in Bavaria a lift has been included in the conversion of a former metalworking shop into an energy-efficient family home (p. 82, Figs. 21–24). This investment, which is anything but standard, does, however, correspond to the interior concept. The existing building with two proper storeys and an 8 m high attic storey under the roof has a relatively small footprint. Once the client had decided to retain the existing building and "extend" it, the architect devised a concept that spreads family life over four storeys but at the same time exploits the qualities of this location. In order to achieve living areas with good illumi-

Maison à Bordeaux (F), 1998, Rem Koolhaas
12 Section, scale 1:400
13 Plan of upper floor, scale 1:400
  a Children's bathroom
  b Child's room
  c Inner courtyard
  d Storage
  e Parents' bathroom
  f Wife's bedroom
  g Husband's bedroom
  h Elevating platform
  i Terrace
14 Plan of ground floor, scale 1:400
  j Terrace
  k Living area
  l Office
15 Elevating platform which allows the wheelchair-bound client to reach all three levels of the house with ease

Low-energy house in Gstadt (D), 2004, Florian Höfer
16 Section, scale 1:400
17 Plan of upper floor, scale 1:400
  d Bathroom
  f Ramp, 10 % gradient
  g Living/dining area
  h Kitchen
  i Study
  j Pantry
18 Plan of ground floor, scale 1:400
  a Bedroom
  b Child's room
  c Play area
  d Bathroom
  e Garage
  f Ramp, 10 % gradient
19 A dog-leg ramp links the two floors; there is also a lift.
20 Kitchen units with legroom underneath, and controls and switches at a height suitable for a wheelchair user

Typology
Housing

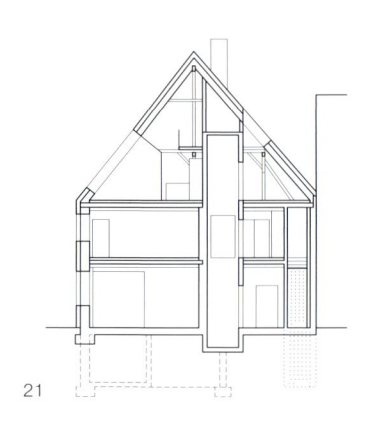

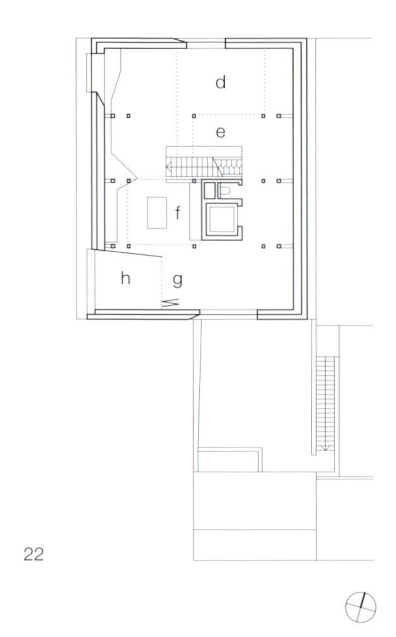

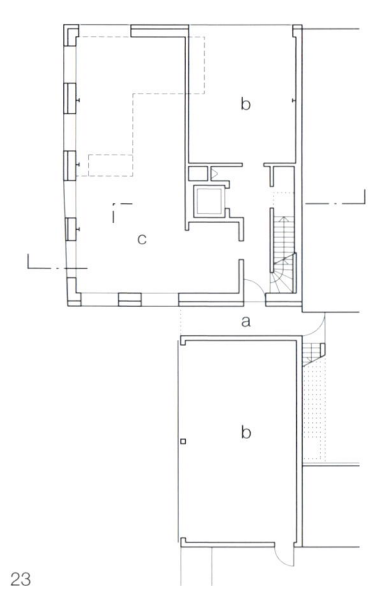

MuUGN House, Memmingen (D), 2008,
Soho Architekten
21 Section, scale 1:400
22 Plan of attic storey, scale 1:400
23 Plan of ground floor, scale 1:400
    a Entrance
    b Garage
    c Workshop
    d Gallery
    e Living room
    f Kitchen
    g Dining room
    h Loggia
24 Exterior view

Solar house for the "Solar Decathlon" competition of the US Ministry of Energy in Washington D.C. (USA), 2007, Darmstadt University
25 Exterior view
26 WC

nation in this heavily built-up town centre setting, the common living area was placed in the attic storey and a gallery built. The bedrooms are on the first floor and the ground floor includes a small workshop. The existing storey heights and purlin roof construction enables a lift to be readily incorporated without major alterations to the building's structure. The lift has proved its worth ever since the building has been in use – not only for guests or looking ahead to old age, but also for the everyday life of a family with two children.

Besides the questions of access, circulation and the associated floor areas, great attention has been paid to the usability of the building. Sensitive points are the fixed elements and facilities plus the sanitary area.
Designing bathrooms and toilets for disabled persons does not mean that a high standard of interior design is impossible. For example, building without barriers is intrinsic to the Darmstadt University project that won the renowned "Solar Decathlon" Award in 2007 (Fig. 26). With new buildings in particular, the concept of building without barriers, if integrated into the planning process at an early stage, often adds nothing to the total cost (Fig. 25).

Terrace houses
In rural districts the once compact, introverted settlement structures are gradually expanding. As freedom from barriers is not an individual phenomenon, but rather a social task, consideration of barrier-free architecture should not end at the plot boundaries, but must include its urban surroundings.
Frequently, structures with more of a rural character react to demographic change by designating the areas for new development on their outer boundaries; the centres of the original settlements age, buildings remain vacant, the existence of small retailers is threatened. The great chances for developing built-up areas, in infill developments and reanimating the buildings and facilities that characterise the centres of cities, towns and villages are not always exploited in such an exemplary way as in Hechtsheim, a suburb of Mainz. Here, houses for nine young families with children have replaced a dilapidated farmstead in the middle of the district (Figs. 29 and 30).

Typology
Housing

27

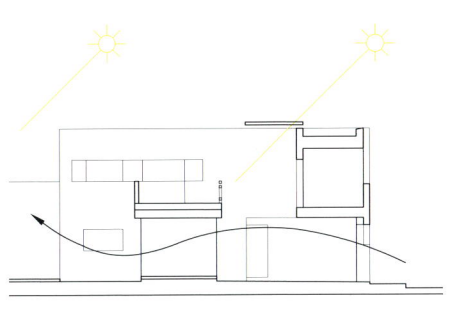

28

The creation of a symbiotic concept has enabled the new buildings to form a sensitive reference to the location, achieved through the imitation and re-interpretation of the courtyard elements typical of this location. At the same time, the overall planning has succeeded in formulating differentiated transitions between public, semi-public and private spaces – architectural answers to the questions of community life that are certainly the equals of the new buildings on the edge of town. The L-shaped ground floors, which contain the common living functions plus the main bedroom, and the other bedrooms above have given rise to nine essentially independent, but nevertheless closely linked, two-storey courtyard houses that together enclose a semi-public square. Following the principle of how the Beguinages were organised, the individual buildings are accessed from this central open space via the private courtyards of the individual houses (Fig. 30, see also "On the history of barrier-free design and construction", pp. 10–11).

Even though the buildings themselves do not correspond to the standardised principles of building without barriers, the influx of, in particular, young families with children has enabled the historic centre of this small Mainz suburb to be revitalised. Totally new neighbourhoods are possible. This is not just a case of supplementing old buildings with new ones; the central area with all its structures and amenities has been strengthened.

Although totally different in terms of its architectural language, the terrace house development in Umbrete in Spain is comparable in terms of its conceptual principles (Fig. 27). In this instance the barrier-free transition between different public areas is almost celebrated. The visitor proceeds from the open street to a semi-public forecourt that provides the access to the houses themselves. The standardised interior layout contains cooking, dining and living areas, three bedrooms, bathroom and WC. The careful organisation of these rooms on different levels not only results in different private external areas, but also room sequences on two levels in some cases. Such designs represent high-quality accommodation for publicly assisted housing (Fig. 28).

29

30

Terrace house development in Umbrete (ES), 2008, Solinas Verd Arquitectos
27 Exterior view of transitions from public street to semi-public forecourt and entrance zone
28 Section through transition sequence

Nine courtyard houses in Mainz-Hechtsheim (D), 2006, Doss & Over
29 The L-shaped housing units form communal courtyards.
30 Location plan

Typology
Housing

Multi-storey housing
When it comes to more condensed housing forms and urban contexts, semi-public, vertical circulation spaces must provide more than just access to apartments. Such staircases in the housing sector rarely attract much attention, so it is astonishing to discover that there are actually some architects who wish to highlight the stairs in a residential building (Fig. 33). The multi-storey housing development in the Zurich suburb of Oerlikon has not only managed to turn functional components into local features in the architectural sense; the staircase has become a semi-public living room in which interaction between residents promotes integration, communication and the formation of communities. The hub of this five-storey apartment block provides access to five apartments on every floor, each with a different interior layout (Fig. 32). It seems almost incidental that this staircase "sculpture" (without set-back risers) could be designed without barriers.

Another good example of experimental multi-storey housing is the "e³" project in Berlin, which takes up the theme of sustainability not purely from the ecological (seven-storey timber structure), but also the economic (prefabrication) and social viewpoints (client group; no barriers) (Figs. 36 and 37).
The particular constructional feature of this project is the seven-storey timber structure, supplemented only by one reinforced concrete staircase with lift and intervening precast concrete bridges. Barrier-free access on all levels is guaranteed. The key elements of the fire protection strategy are relatively short escape routes and a higher overall fire resistance for the construction. The latter is essentially achieved by using larger components for the loadbearing timber structure and enclosing them in incombustible casings.
The client consisted of a group of people, so a form of construction was required that would stabilise the entire building but still allow the users maximum organi-

sational freedom in their interior layouts floor by floor. The loadbearing structure is made from glued laminated timber members erected in the form of a post-and-rail assembly, joined together with steel gusset plates and encased in gypsum fibreboard. This results in floor levels without intervening columns which can be organised to suit individual needs (Fig. 38).

In contrast to professional developers and investors, a client group is interested in solutions tailored to individual requirements rather than standardised forms of construction. Finding and coordinating all those requirements can be very time-consuming, but are intrinsic parts of this process. Nevertheless, it is vital to pose questions regarding communal living in the future, and to find answers. If we define integrative housing as a form of communal co-existence that crosses generation barriers, then the all-embracing nature of a client group, similar to an extended family structure, could be

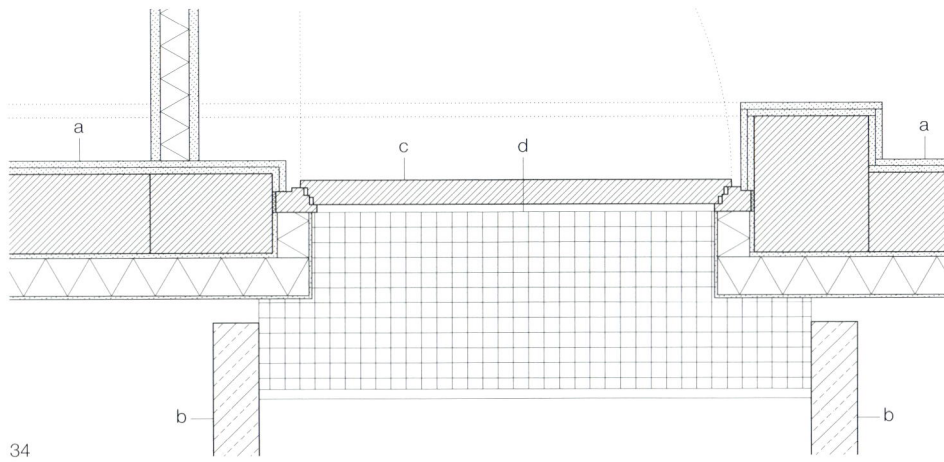

Typology
Housing

34

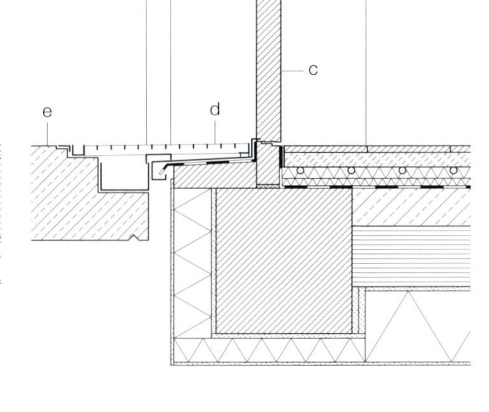

35

a reflection of society. Such self-organised communities that bridge ages, cultures and classes promise a high degree of exchange, integration and robustness compared to mono-functional institutions. We frequently hear the term "multi-generation housing" in this context.

As the "e³" project in Berlin demonstrates, such structures can be reflected in one building complex, a residential development, or throughout a whole district. With all the transformability that must be inherent in these structures, they represent a form of social togetherness that has attracted funds from the Federal Ministry for Family, Senior Citizen, Women & Youth Affairs (BMFSFJ) through the "MehrGenerationenHaus" (multi-generation housing) programme.
Such structures allow the principles of family life and village communities to be learned and then incorporated in planning and organisational structures. The result is settlements where people of all ages can mix quite naturally, learning from and supporting each other. By 2008, after the programme had been in operation for just two years, it was possible to draw up a map with 500 multi-generation housing projects! [4]

In that same year a housing cooperative in Neuss established a new neighbourhood, "Südliche Furth", on the town's former container marshalling yard. This publicly assisted project involves barrier-free accommodation with a gross floor area of about 25 000 m² for more than 700 people (p. 86, Figs. 39–41).
Only 35 of the 255 apartments are privately financed. Even the seven maisonettes satisfy barrier-free criteria. The accommodation is a mix of different apartment forms for young and old. Besides a local centre with common amenities and advice office, there are apartments with between two and six rooms (plus bathroom and kitchen in each case) as well as shared apartments for older persons or those in need

Multi-storey housing in Zurich-Oerlikon (CH), 2007, Graber Pulver Architekten
31 Curving balconies
32 Plan of 1st floor, scale 1:500
    a Living/dining area
    b Room
33 Staircase "sculpture"

"e³" housing project, Berlin (D), 2008, Kaden Klingbeil Architekten
34, 35 Detail of level entrance
        Plan, section
        Scale 1:20
        a Solid timber wall, 316.5 mm
        b Spandrel panel, 115 mm reinforced concrete
        c Entrance door, solid timber
        d Grating
        e Staircase bridge,
           250 mm precast concrete element
36 Section, scale 1:400
37 Seven-storey timber structure
38 Plan of 4th floor, scale 1:400
    f Bridge/entrance
    g Room
    h Kitchen/dining area
    i Living room

36

37

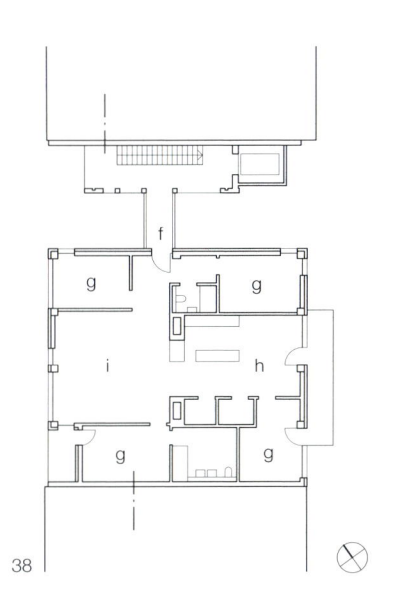

38

85

Typology
Sheltered housing and life as an old person

39

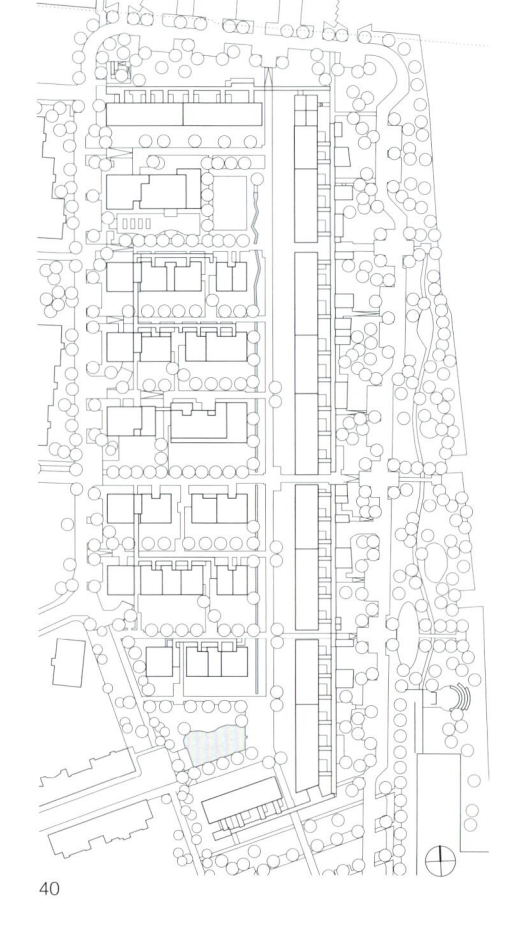

of care. This diverse housing mix offers the chance of cross-generation lifestyles.

Several "layers" screen this new urban addition against railway and traffic noise to the east. The line of four-storey blocks that forms the eastern boundary of the development has noise-attenuating walkways on its eastern side, with only minimal connections to the actual buildings themselves and rolled glass shielding the walkway from the noise sources. Only small outbuildings (as a substitute for the individual cellars normal in Germany) and a few communal open areas are located to the east of this line. Behind the blocks, lower buildings form the urban boundary to the existing housing areas to the west. Pedestrian alleyways, a pond and a water play area plus two squares create a variety of public spaces.
Worries about potential ghettoisation seem to be unfounded. Even though less than 14 % of the apartments are privately financed, the broad mix of different interior layouts and apartment sizes counteracts any stigmatisation. The fact that the different apartment types and sizes are not readily apparent from the outside is certainly advantageous here.
A completely barrier-free design for a project of this size is quite remarkable. Likewise the rigorous planning, which ranges from the configuration of public spaces to access and circulation zones, the building itself, the interior layouts and terrace links right down to the details. It is quite rightly regarded as a pioneering achievement and has been awarded a number of prizes.

Südliche Furth, Neuss (D), 2008, Agirbas Wienstroer
39   Integrated building with sheltered, shared accommodation
40   Location plan, scale 1:3000
41   Detail of level transition between walkway and apartment, scale 1:20

Sheltered housing and life as an old person

The desire for maximum continuity is frequently the determining factor for old people's lifestyles. This is understable because this is a phase of life that is essentially characterised by changes from outside, not self-determined. Those include:
· This establishment of distances to family members, especially children and grandchildren.
· A growing reliance on healthcare facilities, support services and hospital visits.
· The death of a partner.
· The loss of independent mobility.
These factors lead either to the desire to remain self-sufficient within the trusted social and physical environment for as long as possible (which the majority of old people do manage despite the ongoing discussion surrounding support) or to an enforced move to a form of accommodation with less independence but greater security.

The main social, political and economic tasks include:
· The further development of quality-based concepts for various forms of housing and care options.
· Wide availability of marketable everyday aids and care options.
· The spread of, and support for, barrier-free design and construction, and tailored conversions.
· Support for shared accommodation and housing with non-resident carers.
All these tasks must be realised in a form that counteracts any tendency towards ghettoisation and promotes the local- or district-based integration of various generations and cultures.

Alternatives to a continuation of self-sufficient living have continued to develop in

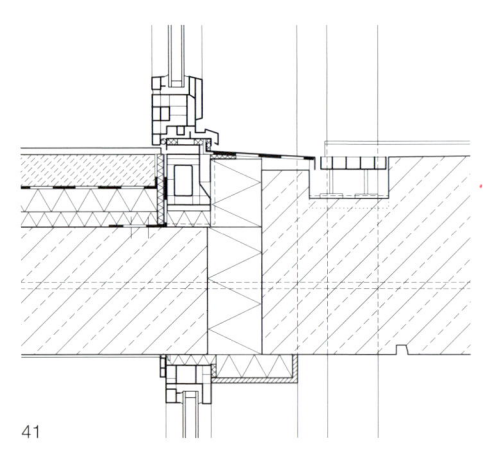

Typology
Sheltered housing and life as an old person

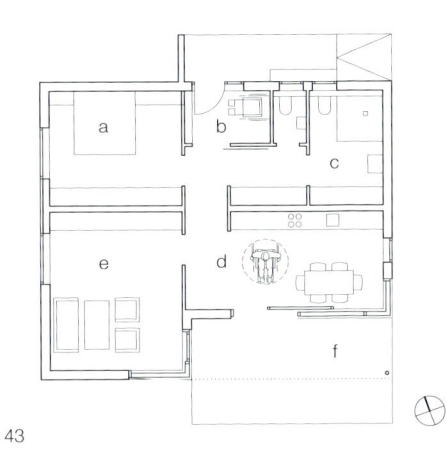

recent years in Germany. This has resulted in a diverse range of options for those unable to rely on their own families. But the various options are not yet available everywhere. However, differentiating the offers for older persons between a "normal" life assisted by non-resident carers plus out-patient service facilities and living permanently in a home with in-patient care services – frequently still the only options today – will in future disappear to some extent owing to the pressures imposed by demographic change.
A fundamental reorganisation of the system for supporting self-sufficient living among older people for as long as possible requires a strengthening of the individual decision-making autonomy. In future older people must be able to decide which type of help they wish to make use of if necessary, regardless of their own abilities. [6]

Memories and the associated social roots play an important role in old age especially. If a relocation is necessary, then the distance (e.g. on foot) to known places and the amenities of the previous neighbourhood should be taken into account. It was proved as long ago as the 1970s that about 80 % of older people who moved no further than a 15-minute walk from their previous home noticed no difference in the number of visits they made or received. This percentage drops to 33 % when the distance is 45 minutes or more. [7]
This is a good reference figure with which it is possible to quantify the relationship with space and positioning in a village community or urban district.

It is recommended that the following amenities be available within a radius of approx. 500 m or a 15-minute walk:
- Shops
- Primary healthcare
- Local public transport
- Café, restaurant
- Barrier-free external amenities, parks, squares, outdoor seating, leisure and recreation areas

We can distinguish between three types of housing for older persons: totally independent accommodation, self-contained dwellings with common facilities, and residential establishments.
Living and caring (for yourself) is seen as a unified whole even if different designs do result in varying spatial relationships, e.g. in residential homes.
Tab. T3 lists the offers and services that may be available.

One example from Hof in Upper Franconia re-interprets the "farm annuity" tradition (Figs. 42 and 43). The pavilion at ground level, with optimised use of energy and designed for wheelchair users, represents an infill development within the garden of a multi-occupancy residential building. The slate-clad, compact, low-energy bungalow with approx. 100 m^2 of floor area is the new home of the building's owner. Besides the design of the bathroom to DIN 18025, the requirements of a wheelchair user are met by the sliding door to the patio and the design of the outward-opening top-hung windows with low-level operating handles (Fig. 42).

T3: Offers and services for individual and communal living schemes [5]

Lifestyle	Everyday social assistance	Care and support services
Individual, self-organised	Individual initiative	Domestic help if required
Individual, with care/assistance	Individual initiative or support by social services	Mobile services organised by care provider if required
Individual, in accommodation for the elderly or sheltered housing, with services	Individual initiative or support by accommodation provider	Help and care, regular advice and consultations plus various additional services organised if necessary
Communal, residential home	Services of accommodation provider	Help and care by accommodation provider
Communal, in neighbourhood form	Individual initiative, communal activities based on neighbourhood initiative	Mutual support, social services if required
Communal, multi-generation housing	Mutual support	Mutual support, social services if required
Communal, shared apartment	Mutual support, arranged by social services, voluntary assistance	Social and care services: self-sustaining, linked with accommodation, decentralised, networked
Communal, household	Mutual support, arranged by social services, voluntary assistance	Social and care services: self-sustaining, linked with accommodation, decentralised, networked

Low-energy house suitable for wheelchair user, Hof (D), 2006, Seeger Ullmann
42 Top-hung window in barrier-free design
43 Plan of ground floor, scale 1:250
  a Bedroom
  b Lobby
  c Bathroom
  d Kitchen
  e Living room
  f Terrace

Typology
Sheltered housing and life as an old person

44

Sheltered housing for senior citizens
Despite supplementary out-patient care options, there will still be a future need for in-patient facilities. But the character of these is already changing and the corresponding units are being downsized. Such changes are evident at the "Haus am Kappelberg" home for the elderly, for instance (Fig. 48). The four storeys with rooms for residents are basically organised as two sheltered housing communities each with a maximum of 14 members (Fig. 47). Despite the size of the building, the colour scheme eliminates any feeling of a sterile, anonymous institution. At the same time, the good contrast between the colours enables good orientation for persons with visual impairments.

The nursing home in Schorndorf uses a much less striking but nevertheless high-contrast colour scheme. The objective behind the design of this nursing home was to create spaces that evoke neither an impression of sterility or anonymity nor that of a clinical institution (Fig. 46). The concept therefore places the individuality and personality of the residents in the foreground. At the same time, it also focuses on the specific needs of older people. Even during a stay in a nursing home, residents should be able to lead a dignified, secure life, in the ideal case feeling completely at home.

The residence options are tailored to the various needs of the residents: secluded in the personal environment of a private room, or communicative, taking part in the life of the home from the comfortable armchairs of the common areas with their homely styling, or semi-privately in the seating alcoves along the corridors, or more actively in the restaurant, in the therapy facilities and out in the fresh air on the terrace (Fig. 44).

No two rooms are alike. Adhering to the idea of individualisation, every room has its own colour scheme. Walls, highlights and curtains in each case form a new combination and hence a different atmosphere. Thanks to their forms and materials, the standard furnishings and fittings are regarded as trusted and ordinary. At the same time, generous storage options and open floor spaces create diverse opportunities for turning the room into a personal home. Each room therefore becomes unique.

45

46

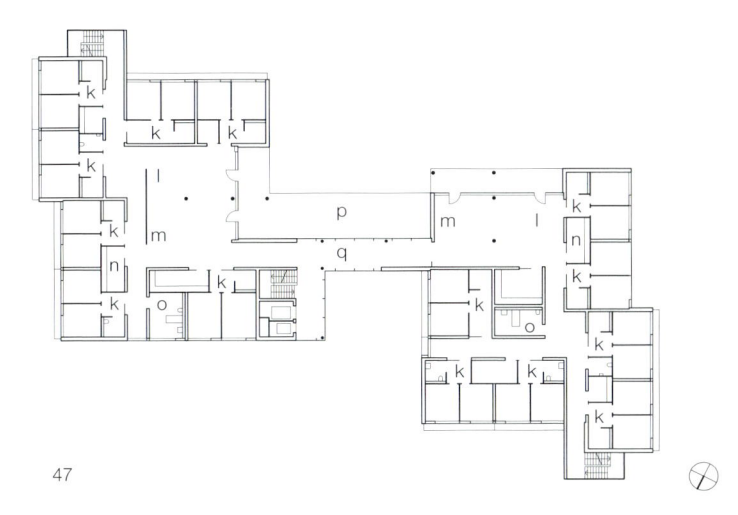

Typology
Sheltered housing and life as an old person

47

The orientation system continues this theme. Instead of abstract room numbers, every room has its own name, e.g. "Castle", "Palace", "Home" – words that certainly suggest "a place to live". In addition, the rooms are arranged alphabetically to ease orientation even further.

Next to every door there is a frame containing the name of the resident plus one or two photos from his or her past. This creates a relationship between the person and the place, a bridge between inside and outside, between yesterday and today.

The main corridor is characterised by the seating groups and pictures the full length of one side. More than 170 picture frames contain collective and individual memories from the lives of the residents: record sleeves, lace doilies, silhouettes, poems, old maps, postcards, posters, photographs, even whole books. In their multiplicity and diversity they stand for the individuality and uniqueness of the lives of the residents. At the same time, this display is constantly changing: residents come and go, items come and go, there is always something new to discover – the wall remains "on the move" (Fig. 45).

Care for persons with dementia
Memories that help to promote a sense of identity, e.g. the corridor wall in the nursing home in Schorndorf, are especially important for persons suffering from Alzheimer's disease. Owing to the gradual loss of the ability to remember, this group of people in particular needs special forms of accommodation, e.g. sheltered housing, grouped flatlets, shared apartments, the organisation of which must be refined and their position better anchored in society.

Constructional concepts for this group of people have been put into practice in the Dementia Competence Centre in Nuremberg. The architects have devised different, themed living atmospheres in the three blocks each with three storeys: bright and modern ("Patio"), dark and secure ("Janus"), and rural traditions ("Country Parlour") (Fig. 52). Each concept has a differentiated colour scheme and characteristic details.

After one year in use it was clear that the residents of the dark, cave-like block settled more quickly than those in the brighter, naturally illuminated one. The bright recess at the entrance to each resident's room, with its patterned wallpaper, has

Nursing home in Schorndorf (D), 2006, Ippolito Fleitz Group
44 Plan of nursing floor, scale 1:800
    a One-bed room
    b Two-bed room
    c Therapy room
    d Terrace
    e Day room
    f Dining
    g Staff
    h Assisted bathroom
    i Reception
45 Guidance system and corridor wall display
46 Day room

Sheltered housing, "Haus am Kappelberg", Fellbach (D), 2007, Wulf & Partner
47 Plan of floors 1–4, scale 1:800
    k Room for two residents with shared bathroom
    l Day room
    m Dining area
    n Staff
    o Assisted bathroom
    p Balcony
    q Connecting corridor
48 Generous amounts of glazing to the common areas linking the two parts of the building
49 Colour scheme

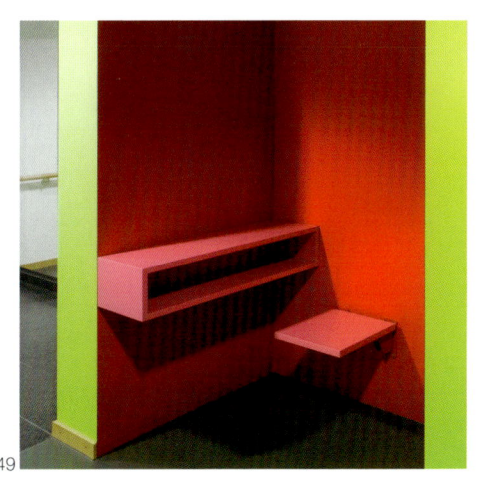

48

49

Typology
Sheltered housing and life as an old person

standard side-hung doors, not the stable doors (i.e. upper and lower halves of each door leaf which can be opened and closed independently) of the rural traditions block. Leaving the upper half open enables a resident to participate in communal life without having to leave his or her room (Figs. 50 and 51).

Care for disabled persons
Organisations such as "Lebenshilfe" care for children, adolescents and adults with mental and multiple disabilities within communities, close to their homes, by way of tailored living, caring and education programmes. "Lebenshilfe" provides comprehensive and extensive care and support for children with learning difficulties and behavioural problems up until primary school age. Parents are also readily integrated into the measures. For those with mental and multiple disabilities, "Lebenshilfe" offers out-patient, semi-in-patient and in-patient facilities from cradle to grave. Holistic care and social integration are central.

As an alternative, the "Living Training Centre" in Aalen, with its three shared apartments which each house four mentally disabled residents, merely offers assistance to achieve a self-determined lifestyle (Fig. 53). Each group organises itself and is only connected via an emergency call system to the on-site care team. During the day, the residents are either employed in a disabled person's workshop or have "normal" jobs. The internal layouts, fully compatible with wheelchair usage, were organised in this new kind of care concept in such a way that the apartments, and hence the groups, can be halved if necessary.

Dementia Competence Centre in Nuremberg (D), 2006, Feddersen Architekten
50 Entrance recess for pair of residents' rooms in "Janus", with stable doors
51 Entrance recess for pair of residents' rooms in "Patio"
52 Plan of all three floors, with "Patio" at the northern end, "Janus" in the middle and "Country Parlour" at the southern end, scale 1:750
    a Assisted bathroom
    b Work room
    c Double room and single room with shared bathroom
    d Single room
    e Laundry
    f Terrace
    g Open-plan kitchen/living area

53 "Lebenshilfe Living Training Centre" in Aalen (D), 2006, VoH Architekten

Typology
Public spaces

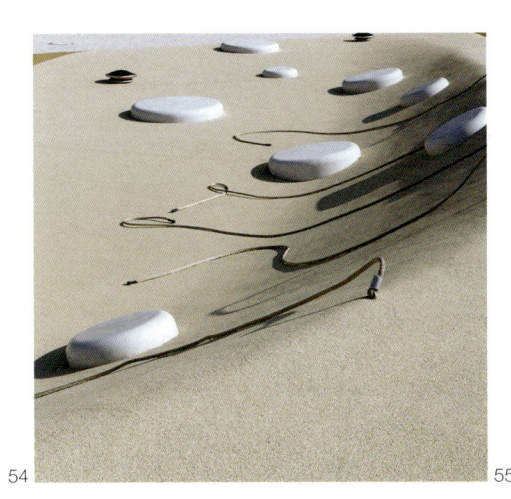

Public spaces

Urban research has established that the assumption of the purely functional urban settlement, as formulated in the Athens Charter, is not sustainable. Detached structures each serving only one usage are ineffective in many respects, unsustainable, offering only limited communicative and integrative options, and the cause of additional traffic. The separation of dwellings and places of work, instead of the creation of mixed-use structures, represents one example that can lead to these negative effects.

There is a growing desire for mixing functions and hence residential areas within easy reach of job opportunities and combined with suitable links to local public transport networks and public amenities such as schools, quality retailing and cultural activities. This has been brought about by the rising cost of mobility and energy, among other factors.

The process of rediscovering the living and working qualities inherent in the centres of our towns and cities, linked with the development of sustainable urban districts and hence public spaces too, increases the incentives to live in such areas. By definition, a public space must be able to be accessed and used by every member of society at any time without any restrictions. It is therefore much more than just an "intervening space". Instead, it enables both nearness and remoteness, is a man-made symbol of our social co-existence.

As long ago as 1920, Theodor Fischer made a clear distinction between the urban functions and intentions of streets and squares. He regarded the constructional formulation of the "right conclusion" to a street corner as an aesthetic requirement "in order to create the impression of the three-dimensional space but with a very particular emphasis on the form of movement" [8]. He envisaged this to be in the form of a square, which invites passers-by to linger and therefore places totally different demands on our senses. The purely aesthetically founded sensory perception of his day has in the meantime become a scientifically founded principle behind a model for conceiving barrier-free urban developments. This structural orientation system assumes that there are prominent constructed orientation points, orientation sectors bounded in space (e.g. districts) and interfaces that supply the basic data for a systematic programme of urban development and orientation. These spatial fundamentals are supplemented at the necessary points by additional orientation aids and guidance systems (see pp. 42–49). [9]

Consequently, public spaces are much more than the multiple-usage intersections of external circulation structures. Arnulf Park in Munich (Fig. 54) and Scharnhauser Park in Ostfildern, a suburb of Stuttgart (Fig. 55), supply good examples of the opportunities and the associated barrier-free usage too. Multi-functional fixtures, which can simultaneously serve as seating, as playthings for small children, as sports apparatus or as a skating course, render possible cross-generation utilisation. Unambiguous routes in defined areas with identifiable enlargements and clear destinations guarantee reliable orientation.

A helpful planning tool in this context – in addition to the design and construction of paths, stairs, ramps, crossings, traffic signals and guidance systems regulated by statutory instruments and standards – is everything that motivates the five senses. Built-in items, borders to spaces, defined spaces and planting are all approaches that stimulate the user according to the

54 Arnulf Park, Munich (D), 2005, realgrün landscape architects
55 Scharnhauser Park, Ostfildern (D), 2001, Janson & Wolfrum

Typology
Public spaces

56

57

58

principle of two senses, foster the perception of the spatial surroundings and promote identification. This also induces chances for emotional links, appreciation and a feeling of responsibility. Much of this becomes possible when the environment can be discovered, experienced and mentally mastered independently. Illumination is especially important for the use of public spaces. Finding the right amount of light to create an atmosphere while offering safety and security is a challenging design task. In addition to the planting and fixtures on Georg-Freundorfer-Platz in Munich, the glancing light creates a non-glare setting that enables this space to be reliably grasped (Fig. 56).

Planned multiple usage makes public spaces more flexible, but at the same time also less specific. They demand a high degree of personal responsibility from their users. This can be seen, for example, in the design of the "stranded rocky iceberg" of the National Opera House in Oslo. Even though, in particular, the open external areas accessible to the public would not fully comply with the standards currently applicable in Germany, this zero-thresholds, zero-balustrades area rising out of the fjord undoubtedly enriches the public life on and around the water in this city. It is not only new land that has been claimed here; the city has been presented with an additional public space (Fig. 59).

Returning public space to the urban environment plays a major role in conversion projects and the search for energy-efficiency potential. Parks in particular are frequently vital providers of fresh air within urban structures and are located in corresponding corridors. An amazing conversion proposal along these lines was inaugurated in New York in 2009. "High Line Park" has been constructed 10 m above street level over a distance of 2.5 km along the tracks of the former "High Line" railway. This is a barrier-free public park accessed via a multitude of stairs and lifts (Figs. 57 and 58). The project shows that efficient design and construction without depleting resources is not just restricted to buildings and structures, but that an urban planning view is necessary every time.

Towns and cities also need places for mourning and reflection. The design of public places of commemoration and remembrance requires great sensitivity. The cemetery in Armea shows the elegance with which such a place can be realised. At the same time, the ramp, with its decidedly aesthetic, hand-crafted design, overcomes the differences in level due to

59

Typology
Public buildings and places of assembly

60

the topography, simplifying the transport of coffins and urns, and making every visit to the cemetery a little easier (Fig. 61).

Public buildings and places of assembly
In the history of European settlements at least, it is the open spaces associated with churches, markets and town halls that have had a decisive influence on the urban landscape. Equally important are the public buildings bordering such spaces, which provide symbols of social order, representation and power, self-assured expositions of the growing strength of the middle classes and a sign of faith.

It is the form and composition of precisely these structures that allow us to assess social behavioural patterns like no other buildings. The reception in Pasing Town Hall is one example in which the completely obvious situation does not initially explain its added value (Fig. 60). There doesn't seem to be anything particularly special about the continuous counter – until we notice that there is legroom underneath across the full width, which means it can be used by wheelchair users at any point.

The reception staff do not sit behind a pane of glass or partly concealed beyond a high counter. Likewise any documents required are easy to see and easy to handle. This is the elimination of barriers and thresholds solved in exemplary fashion on various levels.

Everybody has a right to education, even after finishing school, and this is one of the greatest achievements of our culture. Participation in social and cultural life, learning from this and through this, is simply part of our society, as is clearly shown by the following examples.

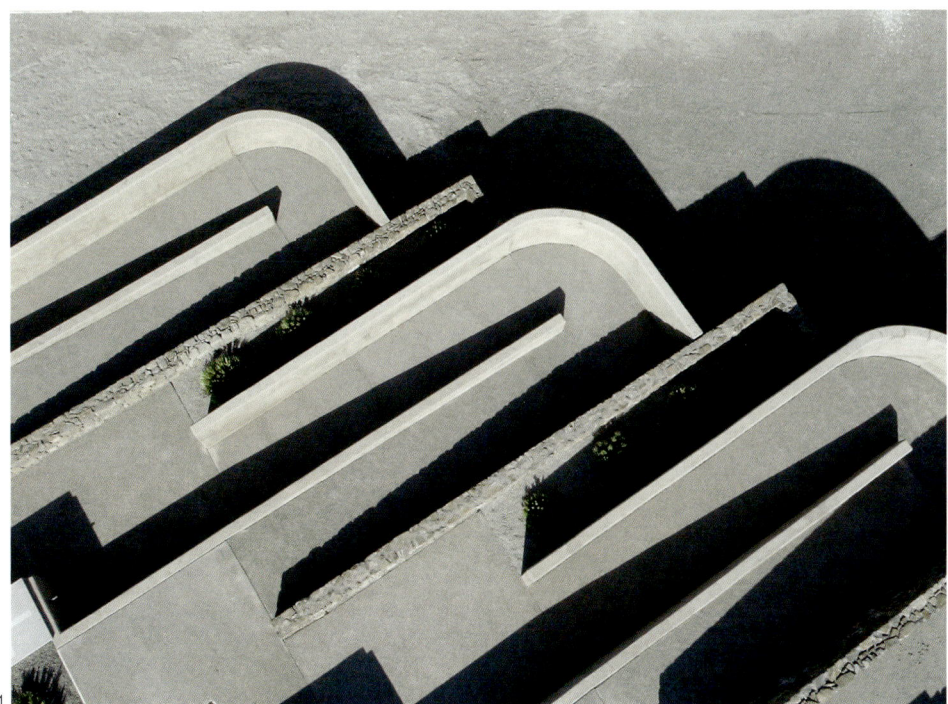

61

56 Lighting at Georg-Freundorfer-Platz, Munich (D), 2002, Levin Monsigny
57, 58 "High Line Park", New York (USA) 2009, Diller Scofidio
59 External amenities, National Opera House, Oslo (NO), 2008, Snøhetta
60 Reception desk, Pasing Town Hall, Munich (D), 2002, Landau Kindelbacher
61 Ramp, Armea Cemetery, Sanremo (I) 2003, Amoretti & Calvi

93

Typology
Public buildings and places of assembly

The entrance to the Georg Schäfer Museum in Schweinfurt is a theatrical one, which all visitors can fully experience equally via the common ramp, which is not just a functional appendage, but rather the real attraction (Fig. 64).
Less spectacular, but with the same self-evident aura, is the new barrier-free ramped entrance (a heritage asset itself) to the historical Nymphenburg Palace in Munich (Fig. 62).

Even if public buildings themselves frequently comply with the standardised rules for building without barriers, this is not always the case with architecture intended to display exhibits. But one remarkable example that proves this doesn't have to be the case is the museum in Manching devoted to Celtic and Roman history. Legroom beneath the fully glazed showcases ensures that wheelchair users can approach as close as non-disabled visitors, to gain the maximum benefit from the exhibits (Fig. 63).

Places of assembly must offer unrestricted usability for all members of the public. The regulations include, for example, dedicated places for wheelchair users within the auditorium (see "Design", p. 56), induction loop systems or supplementary guidance systems. Furthermore, emergency, fire or evacuation scenarios must be given due attention. In such cases large groups of people have to be able to leave the building in the right direction without panicking and as independently as possible. These requirements apply equally to persons with impairments. Corresponding stipulations are laid down in the places of assembly legislation of the federal states (see "Regulatory principles", pp. 26–28, Tab. T5).

It is not only equality of participation in the auditorium that should be considered for all members of society. Where appropriate, it should also be possible for anybody to use the stage or podium. One remarkable example and symbol of a social constitution is the German parliament building in Berlin: the fully height-adjustable lectern is in the form of an upturned L (Fig. 67 shows a similar model). However, the elimination of barriers in this building goes much further. In addition to barrier-free access and circulation, the glass dome directly over the assembly chamber, the highest part of the building, is open to the public and accessed via a ramp.

Typology
Obtaining provisions, shopping

Obtaining provisions, shopping

A self-sufficient lifestyle is characterised by a dwelling's immediate surroundings. Such a lifestyle also includes the ability to take part in social life and, in so far as individual abilities allow, the chance to obtain food, clothing, etc. The ready availability of a range of quality products and the radius of movement within the urban context have a quintessential, sociological and functional significance that is fundamental. The initial integration of these themes should not take place at the *building* planning level, but rather at the *urban* planning level. The origins behind the possibilities for lively, multi-cultural city districts, infill developments and the reanimation of urban centres are to be found in land use planning.

The beginnings of European towns and cities can be traced back to division-of-labour processes, trading and bartering. The marketplaces that ensued have since then provided centres for communication and cultural exchanges. Openness, tolerance and communicative and integrative abilities are at the same time urban assets and features of the elimination of barriers.

One of the most impressive European communication and trading centres is the covered public arcade of the Galleria Vittorio Emmanuele II in Milan (Fig. 66). The following principles for the presentation of goods, which without doubt are still valid today, have all been realised here:
• Accessibility
• Positioning
• Usability
• Goods display areas of maximum size
• Visibility from a multitude of positions and heights
• Connection to public spaces that encourage communication and a long stay
• Naturally lit, covered passages
• The chance to trade in covered spaces regardless of the weather

All these factors are present in modern covered arcades as well, e.g. "Fünf Höfen" in Munich (Fig. 65). Five entrances provide level access to an inner-city block from the pedestrian precinct or roadside footways. The internal courtyards cut into the block provide natural illumination and contact with the "outside world". With glass roofs in some areas, they allow shoppers and passers-by to stroll along in the dry. Cafés in courtyards open to the sky generate a feeling of being in the open air, enable shoppers and visitors to enjoy the rain or the sunshine, and ensure a supply of fresh air.

62 Ramp to visitors centre, Nymphenburg Palace, Munich (D), 2008, Claus & Forster
63 Museum of Celtic & Roman History, Manching (D), 2006, Florian Fischer
64 Ramp to main entrance, Georg Schäfer Museum, Schweinfurt (D), 2000, Volker Staab Architekten
65 "Fünf Höfe", Munich (D), 2003, Herzog & de Meuron
66 Interior view of Galleria Vittorio Emanuele II, Milan (I), 1877, Guiseppe Mengoni
67 Height-adjustable lectern

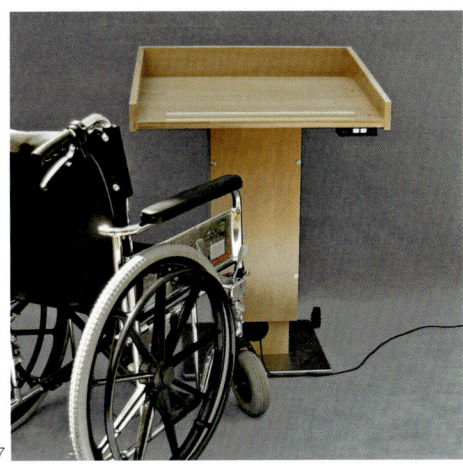

95

Typology
Offices and workplaces

68

Differentiated designs for shopping areas and courtyards using different materials, colours, and various planting schemes (even hanging from the ceiling!), aid orientation even without direct external reference points.

Markets have always been places for social communication. In rural areas this function was taken up and continued by small-scale retailers, although these are now vanishing due to the pressure from cash-and-carry supermarkets, which seldom satisfy social needs. But the Austrian "M-Preis" supermarkets represent a contrast to this general trend (Figs. 68–70). The premises of this chain are designed to suit current market demands, to fit in with respective local structures and to offer a functioning social co-existence, with good-quality buildings. Always accompanied by cafés, parasols and seating (in the car park as well), these are barrier-free shopping experiences and centres of communication and integration for the local community.

Inside each supermarket, 80 cm is the maximum height specified for counters, service facilities and sales tables, which makes everything readily accessible to wheelchair users. Where several identical facilities are available, at least one element must be provided at this height and with legroom underneath. As with a wash-basin, space for legs and knees should be 55 cm deep over a height of at least 67 cm. Here again, a stepped arrangement is possible (the feet need less space). Normal furniture designs are often not fully satisfactory for wheelchair users or persons with mobility impairments. Designing for different levels in front of and behind counters can satisfy these requirements.

Offices and workplaces

One of the prerequisites for leading a self-determined lifestyle is the opportunity to earn a living. This is possible for disabled persons although their impairments may place limitations on their choice of work. However, physical impairments do not present any handicap whatsoever for the activities that are usually necessary in the services sector, for example.

Office structures that do not require rigid room concepts are gradually becoming the norm. As a consequence of this, furniture and fittings are becoming lighter, more mobile, and new ways of working, often arranged to suit individuals, result in the increased flexibility of working spaces. Both traditional office structures and more open-plan variants require sufficient space if they are to be classed as barrier-free, e.g. a manoeuvring area of 150 × 150 cm behind a desk. Office cupboards are sized according to the heights of lever-arch files, but the reach of disabled persons, which lies between 40 and 140 cm, must be considered here. Wheelchair users appreciate a 30 cm high plinth that will accommodate their footrests. Side-hung doors to cupboards

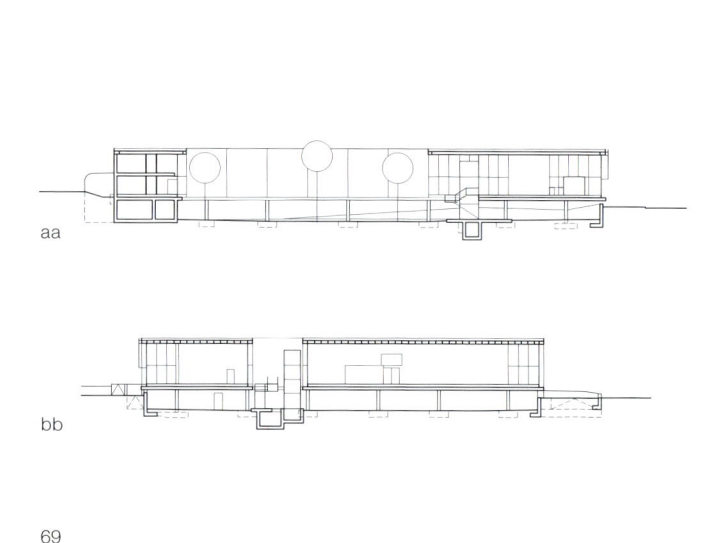

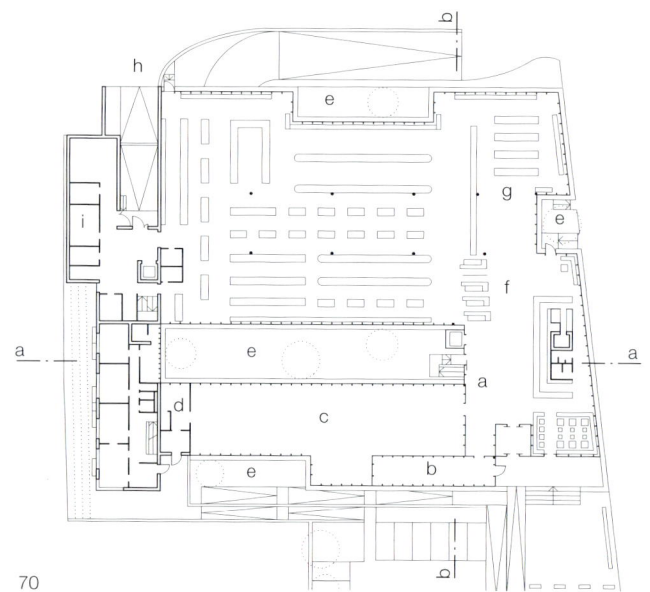

69

70

Typology
Offices and workplaces

71

have proved to be something of a hindrance because of the space they require. Besides the need for identification, accessibility and usability of all permanent facilities as laid down in the standards, the principles of ergonomic and barrier-free designs for desks, office chairs, mobile pedestals, office doors, etc. must be taken into account. It is also important to provide even, non-glare illumination without disturbing shadows or backlighting – for both staff and visitors. This condition applies not only to offices, but also in noisy production areas so that persons with hearing impairments not only have a better view of the workpieces themselves, but can also gather some information from the faces of trainers, colleagues, etc. Optimised lighting is always a benefit to all visually impaired persons. The acoustic working conditions must be considered in offices, especially larger organisational structures, and in conference and seminar rooms. Good sound insulation guarantees functioning communication and legibility of speech, screens off outside noise and improves room acoustics.

According to cl. 71 of Germany's Social Code (SGB IX), every private- and public-sector employer with more than 20 employees must allocate at least 5 % of these jobs to severely disabled people. Where such employers do not employ the prescribed number of severely disabled people, they must pay a levy instead as compensation (SGB IX cl. 77, para. 1). In addition, SGB IX cl. 83 states that every employer is obliged, within the scope of an "operational integration agreement", to implement measures regarding the following factors:
• Appropriate consideration of severely disabled persons in job recruitment
• Attainment of the disabled persons employment quota

• Part-time working practices
• Training of young disabled persons
• Preventive healthcare
• Availability of a works doctor

"Supportive employment" is a state-run scheme designed to help meet these obligations. It provides opportunities for people who as a result of their impairment, e.g. extreme learning difficulties, might not be able to cope with the demands of job training or preparatory measures, but nevertheless do not require the special facilities of a workshop for disabled persons. This scheme is based on the principle of first finding a potential job for the person concerned and then carrying out training measures tailored to that person's needs. On-the-job training must be accompanied by interdisciplinary learning content and key qualifications plus activities for ongoing personal development. As a rule, these two-year individual job training programmes are financed by the Federal Employment Agency. The aim is to integrate a person into society through a proper employer-employee contract which includes the payment of tax and social security contributions.

Workshops for disabled persons are important for those who are unable or not yet able to obtain a job on the open market. Acknowledged workshops offer a wide range of training courses and activities that can be individually adapted to capabilities and possibilities, skills and specialisations.

The "Lebenshilfe" Learning & Training Centre in Ingolstadt is designed according to this system. It has workshops for woodworking, metalworking, textiles, assembly and domestic activities as well as facilities for training courses and events. The building with a floor area

totalling 2400 m^2 is arranged around a spacious, glazed, two-storey courtyard – an "oasis of tranquillity" in the words of the architects – on to which the canteen faces. The different functions of the basic and advanced training facilities for the staff plus the classrooms and workshops for the trainees are all located beneath a long overhanging roof. The perforated metal louvres ensure privacy but still permit a view of the outside world. They also function as sunshades and as an identifying feature. On the whole, the colour scheme with its distinct, warm colours, the short distances between training rooms and workshop areas plus the quiet, concentrated atmosphere is very beneficial for the users.

[1] Stelzer 2009
[2] see VDI 6000 Blatt 1
[3] see VDI 6000 Blatt 6
[4] BMFSFJ/von der Leyen, 2008, pp. 2–3
[5] see also Dettbarn-Reggentin/Reichenbach, 2007
[6] Krämer et al., 2005, pp. 164ff.
[7] Hugues, 1975, pp. 35ff.
[8] Fischer, 1920, pp. 26ff.
[9] Metlitzky/Engelhardt, 2008, pp. 19ff.

Supermarket, Wattens (A), 2003, Dominique Perrault, with RPM Architekten
68 Exterior view
69 Sections, scale 1:1000
70 Plan of ground floor, scale 1:1000
    a Foyer with café
    b Bank
    c Textiles retailer
    d Staff facilities
    e Internal courtyard with planting
    f Checkouts
    g Fruit and vegetables
    h Deliveries
    i Cold store

71 Textiles training area, "Lebenshilfe" Learning & Training Centre, Ingolstadt (D), 2004, Diezinger & Kramer

97

"Résidence de la Rive" nursing home, Onex

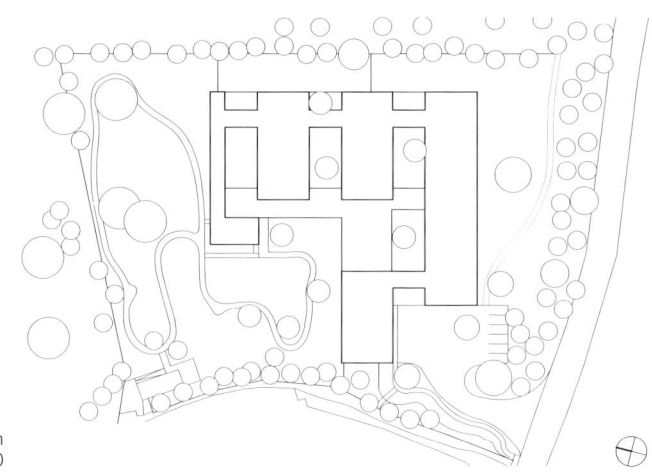

Architects: Atelier Bonnet, Geneva
Structural engineers: Ott & Uldry, Thonex
Date of completion: 2007

Location plan
Scale 1:2000

Building for people whose abilities to perceive their environment are limited can certainly be counted among the most demanding tasks of planning. The learning abilities and powers of remembrance of patients suffering from Alzheimer's disease or dementia decrease constantly – and hence their orientation capabilities, too. Our current state of scientific knowledge does not permit a cure, only a slowing of the process. The spatial environment for these people, reliant on the help of others, is important because the loss of orientation capabilities and power of remembering inevitably leads to uncertainty and anxiety.

The design and construction of an appropriate nursing home therefore forced an in-depth examination of clearly structured, comprehensible organisational principles and opportunities to encourage group dynamic processes on the one hand, and the formation of areas for individual retreat on the other. After seven years of intensive design and construction work, Atelier Bonnet managed to solve this task in a completely natural way in the shape of the "Résidence de la Rive" nursing home near Geneva.

Four residential units are grouped around a north-south circulation zone, creating four intimate, enclosed courtyards framed by the ground-floor reception area. The ground floor is simultaneously circulation zone and common area, with the café as the focal point. A fifth unit on the upper floor above the offices provides options for caring for patients temporarily when their relatives need to be relieved. Despite the great clarity of the internal layout, it was still possible to create areas for withdrawal and retreat. This zoning plan enables users to participate intensively in group life or merely observe it. Stairs and lifts provide access to the individual rooms on the first floor.

Examples
"Résidence de la Rive" nursing home, Onex

Distributing such an establishment over two floors is a non-standard concept. The reason for this approach is to be found in the limited size of the plot and the organizational principle of the nursing concept, which leads from public to semi-public to private rooms. However, this facility does demand more intensive supervision and care services from the personnel than would be the case in groups organised on one floor only. In the form realised here, the link with the semi-public gardens via large opening lights demonstrates both courage and at the same time great sensitivity towards and respect for the needs of the users.

Covered terraces connect the sheltered exterior with the interior; they allow each group to use its own enclosed courtyard. It is said that many Alzheimer's patients have a desire for exercise. An internal layout and routing that essentially permits this in a safe form therefore contributes to their well-being. Plants can also be used to good effect: their changing appearances and fragrances over the course of a year, the possibility of touching them and even becoming involved in gardening work are all factors that support current therapies. Autonomous use of their surroundings and self-determined wayfinding, represents a challenge for dementia patients. The choice of materials and colours for their built environment therefore takes on a special meaning. The interior design of this nursing home shows just how subtly and nevertheless clearly light/dark contrasts can alternate and hence ease orientation in three-dimensional space.

1   Gardens: Access to both the courtyards and even the paths through the gardens are very helpful when it comes to satisfying the urge for movement that can affect many patients. The planting, which reflects the changing of the seasons and stimulates several senses, is robust enough to withstand touching and picking by the residents, who might even carry out some of the gardening work under supervision, to support their therapy.

Section · Plan
Scale 1:750

2	Store	9	Staff room
3	Enclosed courtyard	10	Café
4	Aquarium	11	Reception
5	Multi-purpose room	12	Secretariat
6	Day room	13	Meeting room
7	Workshop	14	Tea kitchen
8	Lounge	15	Office

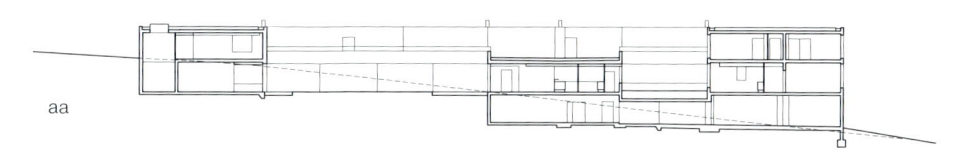

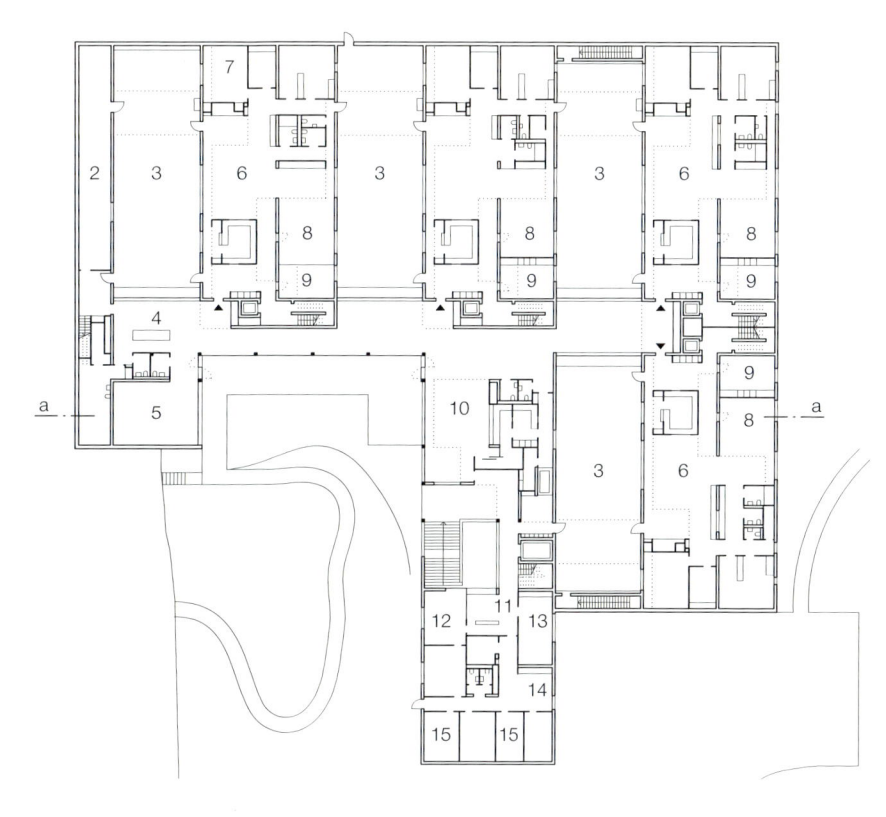

Examples
"Résidence de la Rive" nursing home, Onex

1 Terrace:
The special design of the balustrades alongside the terrace on the upper floor, the panels of which are only approx. 60 cm above FFL, allows the external facilities on the different floors to be used simultaneously. They are transparent and so, on the ground floor, they permit a view through while standing, sitting or even lying down, but nevertheless ensure the necessary standard of safety. At the same time, they break up the height of the enclosing walls to the courtyard in a beneficial way for the users of the ground floor in particular.

Vertical section   Scale 1:10
2 Laminated safety glass,
  2 No. 7.5 mm
3 Wall:
  30 mm render
  160 mm EPS thermal insulation
  130 mm reinforced concrete
  waterproofing, PE sheet
  60 mm XPS thermal insulation
  3 mm sheet aluminium
4 Roof:
  120 mm concrete flags
  separating layer, PE sheet
  120 mm XPS thermal insulation
  (vapour barrier) waterproofing
  300–340 mm reinforced concrete
  10 mm gypsum plaster
5 Laminated safety glass, 2 No. 10 mm
6 Floor:
  5 mm wood-block flooring
  85 mm screed
  separating layer, PE sheet
  40 mm XPS thermal insulation
  20 mm impact sound insulation
  280 mm reinforced concrete
  100 mm lean concrete blinding

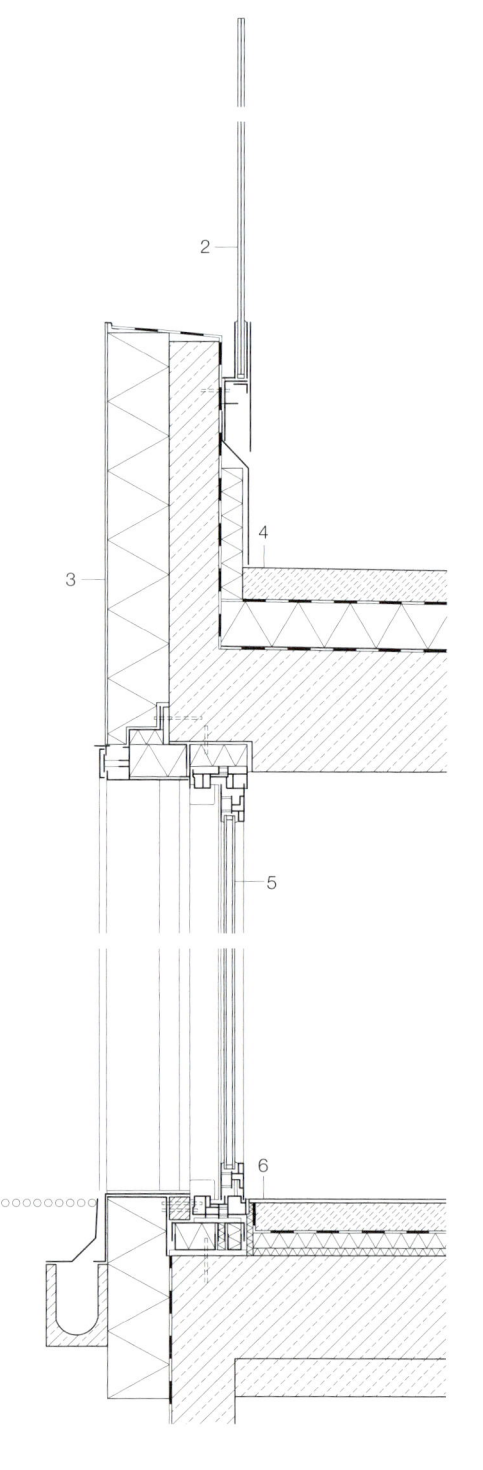

Institute for blind children, Regensburg

Architects:	Georg • Scheel • Wetzel Architekten, Berlin
Project team:	Martina Betzold, Florian Gayer-Lesti, Joao Goncalo Pereira, Nico Kraneburg, Katharina Nailis, Katja Wemhöner
Structural engineers:	ifb Frohloff Staffa Kühl Ecker, Berlin
Date of completion:	2006

This new institute for the blind has been erected in the vicinity of a historical monastery complex in a charming rural landscape on the edge of a new development. The new buildings house a school, a day centre and a residential home for children with visual and multiple impairments. One special challenge faced by the designers was how to deal with the significant slope of the site towards the south-east because the children rely on a stepless meshing of the various rooms. The main corridor therefore includes ramps so that it can track the slope. The single-storey school wings, which are stepped to follow the slope, lead off from this two-storey building which frames one side of the entrance courtyard and contains offices and therapy rooms on its upper floor. The outcome is diverse atmospheres appropriate to the particular usage.

An enclosed forecourt/playground at a higher level forms the urban foyer to the school. Green-grey coal-fired facing brickwork is the material that dominates this, the public side of the building. Contrasting with this, the classroom areas facing the intimate courtyards between the school wings are in timber. These much quieter gardens, also linked by ramps, are open on the narrow side facing the open landscape and, in the background further down the slope, the historical monastery complex.

The internal organisation is determined by a clear orientation system which owing to the severe visual impairments of the children makes use of hierarchical routing relationships. The main corridor leads up the slope via ramps and the corridors leading to the classrooms branch off from the level sections between the ramps. The experience of the particular topography of this site is therefore reflected in the building itself and at the same time becomes the principal organisational criterion.

It is therefore possible to guarantee that the wings on the slope can be reached without the need for stairs. In the building itself, the multi-storey sections are

Examples
Institute for blind children, Regensburg

accessed via three lifts and staircases. The design of the latter employs risers, contrasting nosings and handrails on both sides in order to make them easier to use for children with severe visual impairments. Handrails in the corridors (at two different heights) are very helpful orientation aids.

Outside, grooved surfaces are used for guidance, and raised flat-topped studs ("blisters") mark the boundaries of the playing areas in the gymnasium, for example.

All these aids are so cleverly and so naturally integrated that the architecture itself, through the judicious choice of colours and materials, is adequate as a guidance system within the building; an additional, dominant system is unnecessary. Stone floors and dark wooden door and window frames stand out against the light-coloured walls and ceilings and therefore form a reliable means of orientation. Only adjacent to the lifts are the floor numbers written clearly in Braille on the walls.

Artistically designed, practical elements provide the necessary identification where different internal zones meet. For instance, the children are accompanied by an abacus in the corridor.

Intentionally dissimilar absorptive designs enable users to establish unambiguously which type of room they are in. The quietest form, using materials with a high degree of attenuation, is found in the classrooms. The choice of material – inside and outside – results in both tactile and acoustic guidelines that underscore the special features of the site and integrate the themes of building without barriers in a completely natural way.

Location plan   Scale 1:2000

1. Events
2. Entrance/rest area
3. Multi-purpose room
4. Rhythm room
5. Lobby
6. Music/rhythm room
7. IT room
8. Garden court
9. Sports hall
10. Therapy pool

Examples
Institute for blind children, Regensburg

Section
Part plan of ground floor
Scale 1:1000

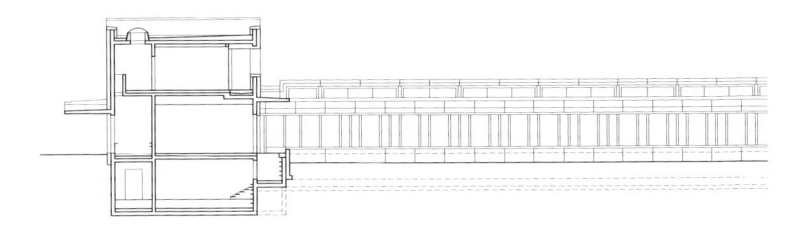

aa

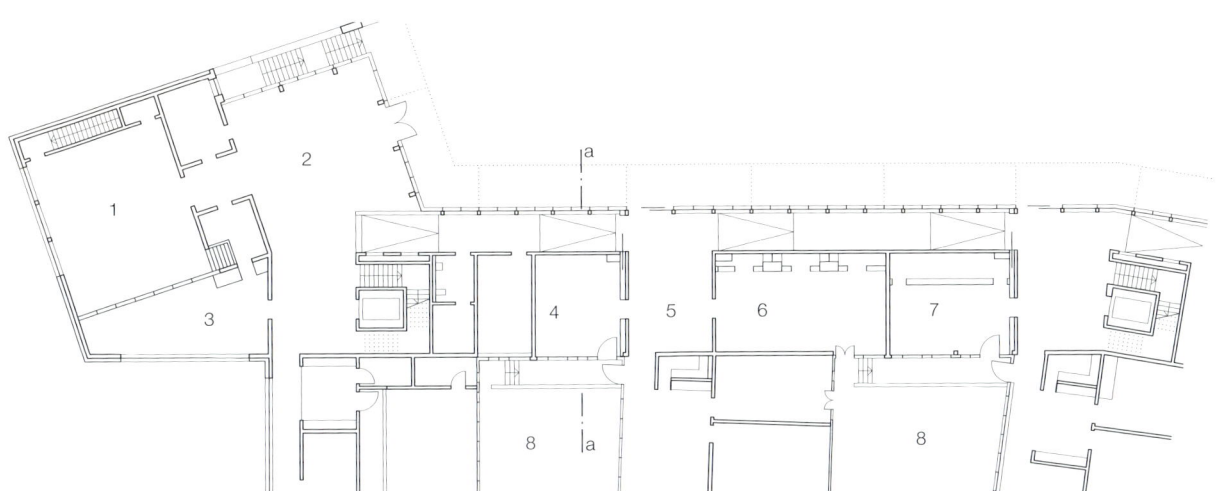

"City Lounge", St. Gallen

Art/architecture: Pipilotti Rist and Carlos Martinez, Widnau
Date of completion: 2005

How an impairment or disability really affects the life of the person concerned can only be assessed by that person. In irritating moments when we are deprived of the full power of our senses it is possible to demonstrate how utterly dependent we are on those five senses. Just how unsettled human perception can become – also for people without impairments – and how that affects the use of our surroundings when they are not arranged in the customary form is demonstrated by the "City Lounge" public space project in St. Gallen. Stepping on to the signal red, unusually soft street surface made from rubber granulate, it becomes clear that a lack of thresholds can have an irritating effect, too. On the one hand, the brightly coloured surfaces of the public space represent a bonding ingredient, on the other they aggravate our orientation. Our perception of the Raiffeisen district as a previously assorted conglomerate of leftover plots and traffic functions has consequently been permanently altered because the new surface combines all the urban spaces into a homogeneous whole extending as far as the pedestrian precinct. Considered as a homely carpet with an agreeable feel, this street surfacing provides the area with an identity and creates the basis for an inviting atmosphere. The idea of the "lounge" open to the public therefore forms the actual leitmotif, creating a harmonious and uniform character. But it also gives this corner of the city, not far from the centre, an appropriate lounge-like quality. Seats, tables and other elements rise out of the red rubber surface seemingly at random. It looks as though a red cloth – which despite its durability is soft to the touch – has been draped over all the fixtures. The ratio of interior to exterior appears to have been reversed because the outer facades of the buildings can also be interpreted as interior facades – as wallpaper to the lounge. It is now, if not before, that the design breaks

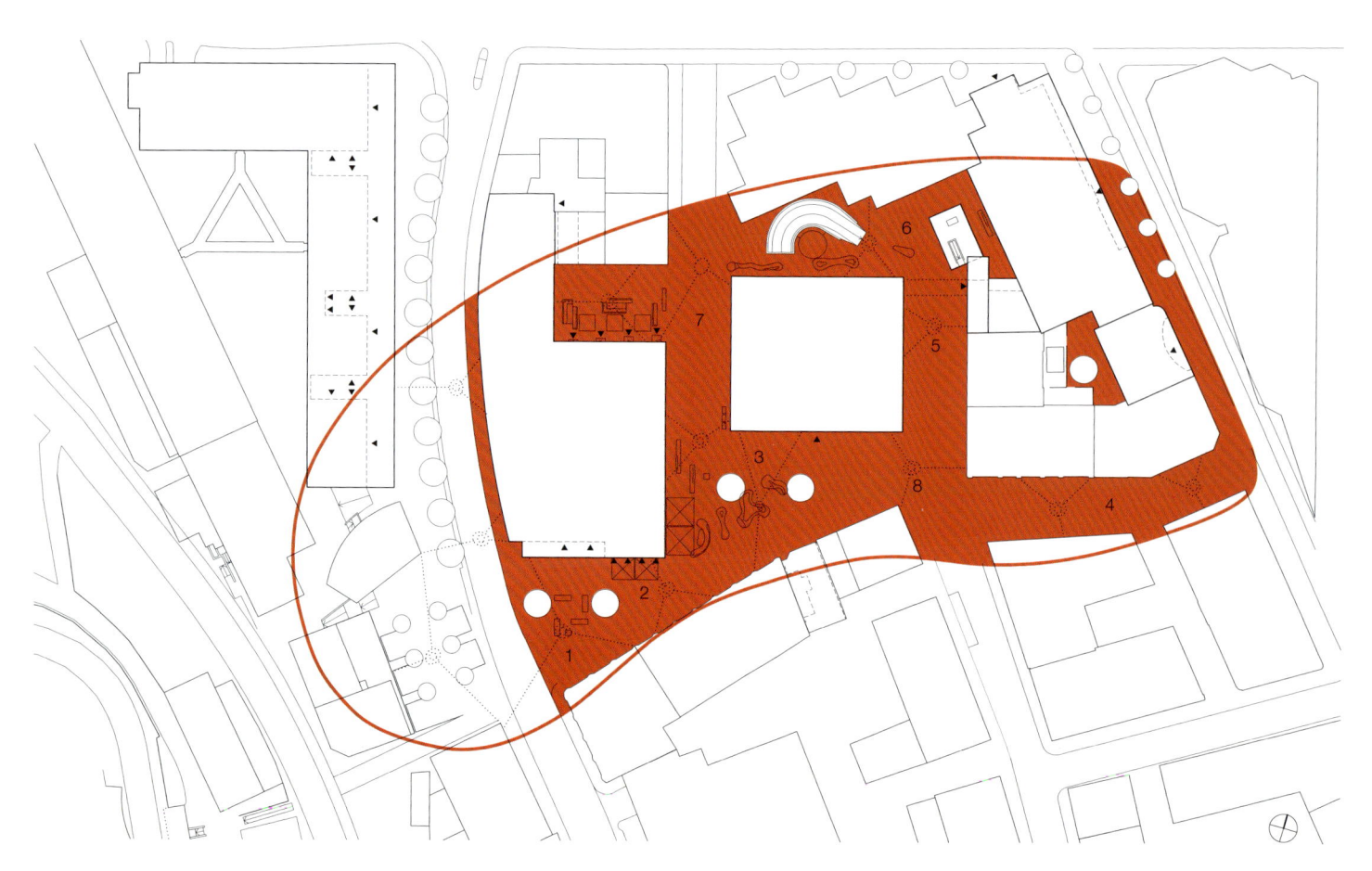

Examples
"City Lounge", St. Gallen

Location plan   Scale 1:1250
Vertical section through seating   Scale 1:10

1   Reception
2   Café
3   Relax Lounge
4   Street Lounge
5   Synagogue
6   Sculptures park
7   Business Lounge
8   "Bubbles" street lighting
9   Covering to seat surface,
     15–20 mm rubber granulate
10   Furniture surface roughened or
     sandblasted
11   Fixing angle, 60 × 60 mm
12   Furniture body, concrete
     or EPS with GFRP covering
13   Asphalt

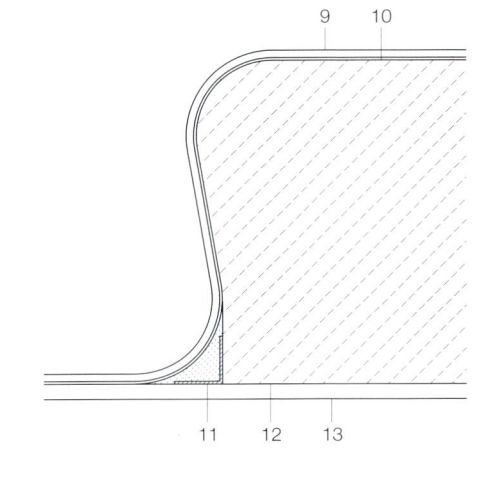

with the accustomed images of public spaces, which are mostly limited to the legible categories of street, square and park, and blurs those of the homogenous ground covering. In this context, the diffuse boundaries between paths and roads and the significance of the surface act as a psychological brake for car drivers because the awareness of the speed of the car increases and more attention must be paid to pedestrians. Slowing down the flows through sensitisation of the car drivers is in turn a vital principle for raising the quality of this district. Modified and hence alien road signs likewise fit neatly into the concept of subversion, just like the rubber granulate-covered car, which makes one parking space "unusable" because it is permanently occupied. The space here is therefore boldly reinterpreted because this particular car invites passers-by to linger, sit down, even lie down. The once mobile object has become immobile furniture because the car, the dominating element in the rest of the urban scene, has in this case been "swept under the carpet"! Projects like this are certainly also valuable because they can be interpreted in many different ways. The public space has been lined with a red carpet leading to a stage; every user is simultaneously spectator and performer. After the first irritating, distracting moments have passed in this wholesale barrier-free environment, our movements and perception demand enhanced concentration. In this project, playing with our senses leads to a mixture of slight intoxication and loss of orientation, which forces us to experience the apparently everyday surroundings with greater sensitivity.

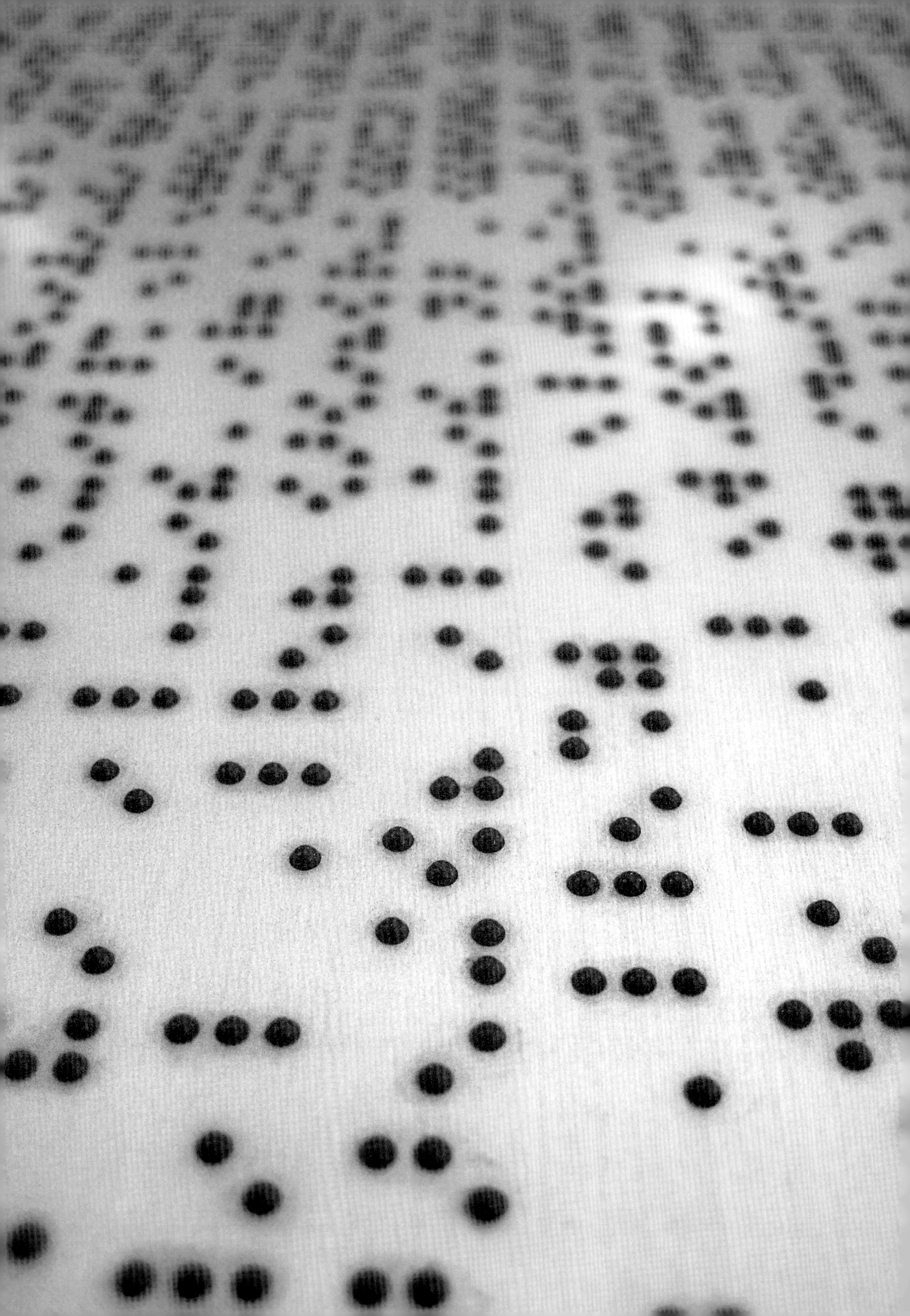

Glossary

Audio installation
In places of assembly and buildings accessible to the public, additional systems should be planned in addition to a PA system so that the hearing impaired can use the building. Three different technologies are available to suit different purposes: → audio induction loops, FM systems, infrared systems.

Audio induction loop
A loop of cable is laid in the floor of the room fitted with an → audio installation. The cable generates a weak magnetic field which enables speech and music signals to be picked up by any person wearing a hearing aid (via the telecoil). No additional receivers are necessary, as would be the case with an FM installation. Also known as audio-frequency induction loop (AFIL).

Barrier-free lift (US: elevator)
The minimum size of the lift car is as follows: clear width 110 cm, clear depth 140 cm. It should be fitted with a horizontal control panel, grab rails, a mirror opposite the door and, in the case of more than two stops, additional audible floor announcements.

Braille
A tactile form of writing, named after Louis Braille, that is based on a 2 × 3 pattern of raised dots (see figure opposite).

Change in level
This is an obstacle that prevents unobstructed horizontal access. Inside a building or other facility a difference in height > 2 cm is classed as an obstacle, outdoors (streets, paths and open spaces) > 3 cm.

Cognitive impairment
Neurological or mental disorders can lead to limited perception skills, diminished recognition, understanding and responses, diminished coordination and orientation abilities, also speech disorders.

Delineator strip
A continuous narrow area dividing different traffic surfaces at the same level, e.g. between cycle track and footway, which provides a tactile and visual contrast between the adjacent surface finishes.

Dementia
A generic term for disorders that involve a loss of cognitive skills such as thinking, remembering, orientation and coordinated actions. A person afflicted by such a disorder can no longer carry out daily activities independently.

Flat-topped bars/studs
A surface finish that ensures a high → tactile, acoustic and visual contrast between itself and the adjacent surface finish.

Gradient
The inclination of a surface, e.g. of a footway in the longitudinal direction. This gradient should not exceed 3 %. With a gradient of 3–6 %, → resting places should be included at regular intervals.

Guidance path surface
A tactile surface consisting of rows of → flat-topped bars/studs that indicate a route and also serve to define a certain area. The use and design of this type of → orientation aid is especially helpful where guiding objects in the form of identifiable edges, e.g. kerbs, house walls, are unavailable. Every measure should be discussed and agreed with the local associations for persons with impairments because they are the ones who understand the local situations with respect to the needs of those affected and can help to integrate such systems uniformly throughout a whole area.

Guidance system
An → orientation aid that eases the use of, in particular, more complex buildings and facilities. Such systems normally rely on a combination of visual, tactile and acoustic information.

Handrail
Handrails 30–45 mm in diameter should be provided on both sides of a stair 85 cm above the pitch line. The inner handrail around the stairwell should not be interrupted, the outer handrail must continue 30 cm beyond the end of the stair at each end of a flight. The legal requirements regarding balustrade heights are not affected by this handrail requirement, which means that a balustrade may need to be constructed separately from the handrail.

Inclusion
This describes the social requirement for every person to be accepted as an individual and for every person to have the opportunity to participate fully in society. Differences and variations are intentionally perceived but their significance diminished or even ignored. In contrast to integration, their existence is neither questioned nor regarded as abnormal by society. The right to participate is socio-ethically justified and relates to all areas of life in which all persons should be able to enjoy unobstructed movement.

Information surface
A tactile surface of → flat-topped bars/studs that indicates particular facilities or equipment. It is generally a square area that indicates the branching of → guidance path surfaces or warns of a → change in level or obstacles.

Kerb (US: curb)
A kerb along one or both sides of a ramp where it is not bounded by a wall or safety barrier prevents the wheels of a wheelchair, usually the small front wheels, from accidentally passing over the edge of the ramp. The top edge of such a kerb should be 10 cm above the finished surface level of the ramp and the kerb should extend over the full length of the ramp if necessary.

Legroom, knee room
In order that a person in a wheelchair can use facilities, e.g. wash-basin, or → operating devices, the clearance underneath must be 67 cm high above floor level.

Motoric impairment
A functional deficiency concerning human muscular-skeletal and locomotor systems caused by damage to the brain, spinal cord, muscles or skeleton, internal organs, limbs or lack thereof.

Movement area, manoeuvring area
A floor space adjacent to a facility or piece of equipment or part of a building which takes into account the special needs of a person reliant on, for example, a wheelchair or mobility aid.

Movement/manoeuvring area for wheelchair user
The floor area necessary for turning a wheelchair, normally 150 × 150 cm.

Multi-generation housing
Residential buildings and facilities that encourage and facilitate cross-generation, neighbourly, everyday involvement between children, youths, adults, persons over 50 and the very old.

Operating devices
Generally buttons, switches, knobs, handles, controls, coin slots, card readers, etc. that require some kind of manual operation.

Orientation aid
Information that helps all persons to use the built environment, especially persons with a → sensorial impairment.

Passing place
An area measuring 180 × 180 cm that, for example, in corridors in public buildings, must be provided every max. 15 m in order to enable two wheelchair users to pass.

Pedestrian refuge
A small section of footway between two carriageways where pedestrians can stop and wait before completing their crossing of a wide road. It should have a → tactile and visually contrasting design so that it is easily located.

Principle of two senses
The simultaneous conveying of information for or by at least two human senses.

Ramp
An inclined surface for connecting areas at different levels. Its → gradient should not exceed 6 %. Where the ramp is more than 6 m long, an intermediate landing min. 150 cm long must be included. The ramp and the intermediate landing must be provided with 10 cm high → kerbs on both sides (where there are no walls or safety barriers) and handrails on both sides, 30–45 mm in diameter, 85 cm above the surface of the ramp. A ramp may not have any → transverse fall.

Reach
A defined horizontal or vertical range that is necessary in order to access an → operating device with the hand and arm.

Resting place
A space or seating area forming part of a footway, staircase or → ramp which has a → tactile and visually contrasting design so that it is easily located.

Sensorial impairment
A deficiency in sensory perception (visual, hearing, lack of sense of smell, taste or touch).

Stepless accessibility
All the levels of a building must be accessible without having to negotiate steps, if necessary via a lift or a → ramp.

Tactile
This concerns the sense of touch and haptic perception. This aspect must be considered for the planning and implementation of orientation and → guidance systems, the provision of information, handrails and → operating devices

Threshold
Entrance thresholds should be avoided. But where they are essential for technical reasons, they may not be more than 2 cm high.

Transverse fall
The inclination of a surface perpendicular to the normal direction of usage. This fall should not normally exceed 2 % on footways, 6 % at access points.

Appendix

Standards, directives, statutory instruments (selection)

Arbeitsstättenverordnung (ArbStättV, Places of Work Act) and Artbeitsstättenrichtlinien (ASR, Places of Work Directives), 12 Aug 2004, last amended on 18 Dec 2008
BauGesetzbuch BauGB
Behindertengleichstellungsgesetz (BGG, Disabled Persons Equality Act), 27 Apr 2002, Federal Gazette I, 2002, p. 1674
DIN 1450 Lettering; legibility
DIN 18024-1 Barrier-free built environment – Streets, squares, paths, public transport, recreation areas and playgrounds – Design principles
DIN 18024-2 Construction of accessible buildings – Publicly accessible buildings and workplaces, design principles
DIN 18025-1 Accessible dwellings – dwellings for wheel chair users, design principles
DIN 18025-2 Accessible dwellings; – design principles
DIN 18030 The work on this standard, the task of which was to develop DIN 18024 and 18025 further, has been abandoned after 10 years because a consensus could not be reached.
DIN 18040-1 (draft standard) Construction of accessible buildings – Design principles – Publicly accessible buildings
DIN 18040-2 (draft standard) Construction of accessible buildings – Design principles – Dwellings
DIN 18040-3 (temporarily shelved) Traffic infrastuctures
DIN 18041 Acoustic quality in small to medium-sized rooms
DIN 18065 Stairs in buildings – Terminology, measuring rules, main dimensions
DIN 32975 Designing visual information in the public area for accessible use
DIN 32981 Special devices for blind and partially sighted persons on traffic signals – Requirements
DIN 32984 Ground surface indicators in public areas
DIN 33402-2 Ergonomics – Human body dimensions – Values
DIN 33402-2 Beiblatt 1 Human body dimensions – Values; Supplement 1: Application of body dimensions in practice
DIN 33404-2 Berichtigung 1 Ergonomics – Human body dimensions – Values, Corrigenda to DIN 33402-2
DIN 33402-3 Human body dimensions; movement room at different normal positions and movements
DIN 66079-4 Graphical symbols for public information – Graphical symbols for disabled persons
DIN 77800 Quality requirements for providers of "Assisted living for the elderly"
DIN EN 81-40 Safety rules for the construction and installation of lifts – Special lifts for the transport of persons and goods – Stairlifts and inclined lifting platforms intended for persons with impaired mobility
DIN EN 81-70 Safety rules for the construction and installations of lifts – Particular applications for passenger and goods passengers lifts – Accessibility to lifts for persons including persons with disability
DIN EN 179 Building hardware – Emergency exit devices operated by a lever handle or push pad, for use on escape routes – Requirements and test methods
DIN EN 1125 Building hardware – Panic exit devices operated by a horizontal bar, for use on escape routes – Requirements and test methods
DIN EN 12183 Manual wheelchairs – Requirements and test methods
DIN EN 12184 Electrically powered wheelchairs, scooters and their chargers – Requirements and test methods
DIN ISO 3864-3 Graphical symbols – Safety colours and safety signs – Design principles for graphical symbols for use in safety signs
ISO 4190-5 Lift (Elevator) installation – Control devices, signals and additional fittings
DIN Technical Report 124 – Products in Design for All Resolution of the Council of the European Union and the representatives of the Governments of the Member States, meeting within the Council of 17 March 2008, on the situation of persons with disabilities in the European Union (2008/C 75/01)
Federal state building regulations
Federal state disabled persons equality legislation
Federal state garages legislation
Federal state places of assembly legislation Grundgesetz (GG, Basic Law for the Federal Republic of Germany), art. 3, para. 3, sent. 1; Act for Amending the Basic Law, 27.10.1994, Federal Gazette I, 1994, p. 3146
Federal state restaurants legislation
Federal state retail premises legislation
Heimgesetz (HeimG, Homes Act), 7 Aug 1974, last amended on 31 Oct 2006
Heimmindestbauverordnung (HeimMindBauV, minimum building regulations for homes for the elderly and nursing homes for adults), 27 Jan 1978, last amended on 25 Nov 2003
ISO 4190-5: Lift (elevator) installation – Control devices, signals and additional fittings
Musterbauordnung (MBO, Model Building Code), Nov 2002
Sozialgesetzbuch (SGB, Social Code), SGB IX – Rehabilitation and participation of disabled persons, 19 Jun 2001, last amended on 31 Dec 2008
United Nations' "Convention on the Rights of Persons with Disabilities"
VDI 6000 Technical rule – Provision and installation of sanitary facilities
- Blatt 1: Private housing
- Blatt 5: Housing for the elderly, old people's homes, nursing homes
- Blatt 6: Kindergarten, day-care centres, schools
VDI 6008 Blatt 1 Technical rule – Barrier-free buildings for living – Standards for electrical installations and lifts

The following is a selection of publications relevant for Austria and Switzerland:

Austria:
Federal legislation
Behindertengleichstellungsrecht (Disabled Persons Equality Act) passed by Austria's National Assembly on 6 Jul 2005 and announced in Federal Gazette I No. 81/2005 and Federal Gazette I No. 82/2005; it consists of the following:
- BundesBehindertengleichstellungsgesetz (BGStG, Federal Disabled Persons Equality Act)
- Diskriminierungsschutz in der Arbeitswelt (protection against discrimination at work) in cl. 7a–7q of Behinderteneinstellungsgesetz (Disabled Persons Employment Act)
- Behindertenanwaltschaft (advocacy for disabled persons) to cl. 13b ff. of Bundesbehindertengesetz (Federal Disabled Persons Act)
- Recognition of the Austrian sign language as an independent language in the Austrian Constitution, art. 8, para. 3

Standards
ÖNORM B 1600 Building without barriers – Design principles
ÖNORM B 1601 Special buildings for handicapped or old persons – Design principles
ÖNORM B 1602 Barrier-free buildings for teaching and training and possible accompanying facilities
ÖNORM B 1603 Barrier-free buildings for tourism – Design principles
ÖNORM B 2457 Chair lift for handicapped persons; design rules
ÖNORM B 2607 Playgrounds – Planning principles
ÖNORM V 2102 Technical aids for visually impaired and blind persons – Tactile ground surface indicators
ÖNORM V 2104 Technical aids for blind, visually and mobility impaired persons – Safety devices for construction and dangerous sites
ÖNORM V 2105 Technical aids for visually impaired and blind persons – Tactile inscriptions and information systems
ÖNORM A 3011 Graphic symbols for public information
ÖNORM A 3012 Visual guiding systems for public information – Orientation supported by directional arrows, graphic symbols, text, light and colour
ÖNORM K 1105 Wheelchairs – Terminology, classification and dimensions

Switzerland:
Federal legislation
Federal Constitution of the Swiss confederation; SR 101, 18 Apr 1999 (position as of 30 Nov 2008), art. 8, "Equality before the law"
Behindertengleichstellungsgesetz (BehiG, Disabled Persons Equality Act), 13 Dec 2002
Behindertengleichstellungsverordnung (BehiV, Disabled Persons Equal Opportunities Act), 19 Nov 2003

Standards
SIA 500 Barrier-free building

Appendix

Bibliography

Bösl, Elsbeth: Konstruktionen von Behinderungen in der bundesdeutschen Behindertenpolitik. In: Arbeit Gender Wissen, 2009

Burckhardt, Jacob: Die Kultur der Renaissance. Essen, 1860

Degenhart, Christine: Generationengerechtes Bauen. Kissing, 2008

Department of the Environment, Transport & the Regions (DETR): Guidance on the use of tactile paving surfaces. London, 1998

Department of the Environment, Transport & the Regions (DETR): Parking for disabled people. London.

Dettbarn-Reggentin, Jürgen; Reichenbach, Michael (ed.): Bau- und Wohnkonzepte für alte und pflegebedürftige Menschen. Praxisbeispiele, Planungshilfen, technische Daten und medizinische Grundlagen. Merching, 2007

Eberwein, Hans; Knauer, Sabine (ed): Integrationspädagogik: Kinder mit und ohne Beeinträchtigung – lernen gemeinsam. Ein Handbuch. Weinheim, 2002

Eco, Umberto (trans. Brock): The Mysterious Flame of Queen Loana. Milan, 2004

Federal Office For Building & Regional Development (BBR); Institut f. Erhaltung & Modernisierung von Bauwerken e.V. (IEMB), Berlin TU (pub.): Wohnen ohne Barrieren – Erhöhte Wohnqualität für alle. Meckenheim, 2008

Federal Ministry for Family, Senior Citizens, Women & Youth Affairs (BMFSFJ); von der Leyen, Ursula (pub): Starke Leistung für jedes Alter – Erste Ergebnisse der Wirkungsforschung im Aktionsprogramm Mehrgenerationenhauser, 05/2008

Fischer, Theodor: Sechs Vorträge über Stadtbaukunst. Extended reprint of 1st ed. (1920), ed. by Matthias Castorph. Munich, 2009

Frejka, Tomas et al.: Childbearing trends and policies in Europe. Max Planck Institute for Demographic Research, Rostock, 2008

Frohnmüller, Sabine: Barrierefreies Bauen – Neue Vorschriften und technische Regeln, bau intern 3I4, 2009, pp. 10–11

Gemeinsamer Fachausschuss f. Umwelt & Verkehr (GFUV, Joint Committee for Environment & Transport), workshop on ground surface indicators; Behling, Klaus: Anforderungen an die Profile und den Einsatz von Bodenindikatoren im öffentlichen Raum. Berlin, 2008

Herwig, Oliver: Universal design. Basel, 2008

Hugues, Theodor: Die altengerechte Wohnung. Munich, 1975

Krämer, Stefan et al.: Wohnen im Alter Entwicklungsbilanz und Perspektiven. In: Wüstenrot Foundation (ed): Wohnen im Alter. Stuttgart, 2005, pp. 152–167

Lanz, Stephan: In Europa mehr Initiative und Kraft entwickeln. In: Multiple City, ed. by Winfried Nerdinger, Sophie Wolfrum. Berlin, 2008, pp. 294–298

Marx, Lothar: Barrierefreiheit als Herausforderung. In: Fischer, Joachim/Meuser, Philipp (ed.): Barrierefreie Architektur: alten- und behindertengerechtes Planen und Bauen im 21. Jahrhundert. Berlin, 2009, pp. 24–35

Mayer, Susanne: Gefährdete Existenzen. In: Die Zeit, 18 Sept 2008

Mederios Kent, Mary; Haub, Carl: The demographic divide: What it is and why it matters. Population Reference Bureau, Washington D.C., 2005

Metlitzky, Nadine; Engelhardt, Lutz: Barrierefrei Städte bauen: Orientierungssysteme im öffentlichen Raum. Stuttgart, 2008

Rau, Ulrike (ed.): Barrierefrei: Bauen für die Zukunft. Berlin, 2008

Rau, Ulrike in: Dies. (ed.): Barrierefrei: Bauen für die Zukunft. Berlin, 2008, pp. 68–75

Secretariat for the Convention on the Rights of Persons with Disabilities: Accessibility for disabled – A design manual for a barrier-free environment. http://www.un.org/esa/socdev/enable/designm/intro.htm

Senior Building Authority in the Bavarian Ministry of the Interior; Bavarian Ministry for Employment & Social Structure, Families' & Women's Affairs; Bavaria Chamber of Architects (pub.): Barrierefreies Bauen 1, Barrierefreie Wohnungen, Planungsgrundlagen. Author: Marx, Lothar. Munich, 1992

Senior Building Authority in the Bavarian Ministry of the Interior; Bavarian Ministry for Employment & Social Structure, Families' & Women's Affairs; Bavaria Chamber of Architects (pub.): Barrierefreies Bauen 2, Öffentlich zugängliche Gebäude und Arbeitsstätten, Planungsgrundlagen. Author: Ebe, Johann et al. Munich, 1999

Senior Building Authority in the Bavarian Ministry of the Interior; Bavarian Ministry for Employment & Social Structure, Families' & Women's Affairs; Bavaria Chamber of Architects (pub.): Barrierefreies Bauen 3, Straßen, Plätze, Wege, Öffentliche Verkehrs- und Grünanlagen sowie Spielplätze, Planungsgrundlagen. Author: Ebe, Johann et al. Munich, 2001

Sport England (pub.): Accessible Sports Facilities. London, 2010

Stadler, Hans; Wilken, Udo: Pädagogik bei Körperbehinderung. Studientexte zur Geschichte der Behindertenpädagogik, vol. 4. Weinheim/Basel/Berlin, 2004

Statistical Offices of the Federation & the Länder (pub.): Bevölkerung Deutschlands bis 2050, 11. koordinierte Bevölkerungsvorausberechnung. Nov 2006

Statistical Offices of the Federation & the Länder (pub.): Demografischer Wandel in Deutschland, Bevölkerungs- und Haushaltsentwicklung im Bund und in den Ländern. Wiesbaden, Dec 2007

Stelzer, Tanja: Ich will doch nur spielen. In: Mein Kind schafft das. Zeit-Magazin No. 32, 30 Jul 2009

United States Access Board: Americans with Disabilities Act and Architectural Barriers Act Accessibility Guidelines. Washington D.C., 2004

WHO: ICIDH-2 (International Classification of Impairments, Activities & Participation). Geneva, 1999

Manufacturers, companies and trade associations (selection)

The manufacturers named in this publication and listed below represent a selection of possible suppliers. Inclusion in this book and/or the list is not to be understood as a recommendation but merely as an example and the authors make no claim as to the completeness of the information.

Baden-Württemberg Chamber of Architects
Danneckerstr. 54
70182 Stuttgart
www.akbw.de

Hesse Chamber of Architects & Urban Planners
Mainzer Str. 10
65185 Wiesbaden
www.akh.de

North Rhine-Westphalia Chamber of Architects
Zollhof 1
40221 Düsseldorf
www.aknw.de

Saxony Chamber of Architects
Goetheallee 37
01309 Dresden
www.aksachsen.org

Saxony-Anhalt Chamber of Architects
Fürstenwall 3
39104 Magdeburg
www.ak-lsa.de

Bavaria Chamber of Architects
Advice Centre for Building without Barriers
Waisenhausstr. 4
80637 Munich
www.byak.de

Bayerische Stiftung für Qualität im Betreuten Wohnen e.V.
Barbarossastr. 19
81667 Munich
www.stiftung-betreutes-wohnen.de

Bundesarbeitsgemeinschaft der Senioren Organisationen (BAGSO) e.V.
Eifelstr. 9
53119 Bonn
www.bagso.de

Bundesinteressenvertretung der Altenheimbewohner e.V. (BIVA)
Vorgebirgsstr. 19
53913 Swisstal
www.biva.de

Schweizerische Fachstelle für behindertengerechtes Bauen
Kernstr. 57
8004 Zurich
www.hindernisfrei-bauen.ch

United States Access Board
1331 F Street, NW, Suite 1000
Washington, DC 20004-1111, USA
www.access-board.gov

National Disability Authority
25 Clyde Road
Dublin 4, Republic of Ireland
www.nda.ie

Office for Disability Issues
PO Box 1556
Wellington, New Zealand
www.odi.govt.nz

Secretariat for the Convention on the Rights of Persons with Disabilities
Two United Nations Plaza, DC2-1372
New York, NY 10017, USA
www.un.org/disabilities

Appendix

Index

access	50ff., 82
- without steps	102
accessibility	41, 82, 97
- for all	16
accommodation for the elderly	87
acoustic information	73
additional living space	73
adventure area	70
aims of DIN 18040	19
alarms	21
alerting	73
all-day schooling	77
Alzheimer's disease	33, 39, 89, 98f.
ambulatory care	13
anthropometric data	35, 79
artificial limb	15
assistance programme	24f.
audio induction loop	37, 57
automatic door	39
automatic machine	39
automatic opening mechanism	50
average size of children	78
backrest	58
balcony	69
balustrade	69
Basic Law for the Federal Republic of Germany	19
bath	67f.
bedroom	65f.
Beguinage	11
birth rate	30f.
Braille	40, 53
building entrance	50f., 72
building for children	75
building for the disabled	15ff.
building with public access	55ff.
bus stop	47f., 70
button	51
car parking space	71
caring programme	90
cash-and-carry supermarket	96
centres for communication and cultural exchanges	95
change in level	43
change of use	56
changing	
- cubicle	72
- table	60
Charter of Fundamental Rights of the European Union	19
childhood	12
circulation zone	51
clear opening	51
cognitive impairments	9
cognitive skills	39
communal household	87
communal living in neighbourhood form	87
communal living	87
compensation	97
compulsory stairs	53f.
contrast	37f.
- negative	38
- positive	38
controls and handles	39f., 71f., 78, 82
conversion project	92
conversion	66
costs	33
counter	61f., 73, 93, 96
crossing	91
crossing place	48
delayed closing feature	52
dementia	33, 89f., 98f.
demographic developments	30f., 41
demographic divide	30
detached house	80f.
difference in brightness	38
dimensions of the human body	35
DIN 18024	18ff.
DIN 18025	18ff., 33
DIN 18030	21
DIN 18040	19, 21, 23
DIN 77800	21
disability	17
distinguishability	41, 97

door	50ff., 71
- automatic	39
- handle	52
- in public building	52
- revolving	51
- side-hung	51
dwelling entrance door	52, 63
education programme	90
educational therapy day centre	78
emergency call system	69
entrance	50
equality legislation	16
escalator	45, 71
Euro-key	57
evacuating	73
extended family living form	13
fall height	46
Federal Building Code	20
federal government	24
federal state building regulations	19, 26ff., 55, 64
Federal state governments	24f.
fire protection	41f.
fitting-out	72
fittings and furnishings	65
flat-topped studs ("blisters")	45
floor finish	
- indoors	71
- outdoors	50, 71
FM system	37
fold-down support rail	59
fragrance garden	47, 99
freedom from barriers	15ff., 26, 73
Fuggerei	11
furniture and fittings	96
General Anti-Discrimination Act	19f.
general services	33
glass door	39
gradient	45, 54
grammar school	76
grooved element	43f., 48
grouped flatlets, shared apartment	13, 90
guidance path surface	44
guidance system	50, 79, 92
- visual	37f.
- tactile	23
hallways in dwellings	65
hand drier	72
handle	51f.
handrail	52f., 64, 102
Haussmann	12
hearing	36
home for the elderly	13f.
home for the totally crippled	14
Homes Act	20
homes for those injured in the war	15
horizontal circulation	51
horizontal lift controls	55
housing for wheelchair users	73
housing tailored to the needs of the elderly	33
housing	63ff., 80ff.
hydraulic jack	47
illumination	92, 97
impairments	
- cognitive	9
- motoric	9, 35f.
- mental	9
- sensorial	9, 35f.
incline	45, 50
inclusion	42
inclusive design	16
individual lifestyle	87
induction loop system	94
induction	36
inductive system	37
industrialisation	13
infill development	83, 95
information surface	43f., 50
infrared system	37
infrastructure	33
in-patient care	87

International Classification of Functioning, Disability and Health (ICF)	9
jamb	52
kerb	46, 54
kitchen	65f., 73
legroom	61, 66, 77, 81, 93, 97
life as an old person	86ff.
life expectancy	32
lift	40, 51ff., 64, 71, 81, 92
lift call button	51
lighting	72
living room	65f.
Living Training Centre	90
longitudinal gradient	50
lowered kerb	49
low-floor technology	47
luminance	38
main	
- entrance	50f.
- path	50, 70
marketplace	95
marking	39
markings on the handrail	53
mechanical ventilation system	68
MehrGenerationenHaus (multi-generation housing)	85
mental impairments	9, 39
Minimum Building Regulations for Homes	21f.
mirror	55, 58, 68, 72
Model Building Code	21ff., 55, 64
Montessori	12
more intensive care services	13
motoric impairments	9, 35f.
motoric issues	35f.
mounting height	39
movement area	43, 63ff., 96, 70
moving walkway	38
multi-generation housing	85, 87
multi-storey housing	84ff.
negative contrast	38
no change in level	69, 87
number of households	32
nursing home	13f., 82f., 98f.
obtaining provisions	95f.
OECD	77
office	96f.
open space	42ff., 105
operational integration agreement	97
orientation system	45, 72, 79, 89
outdoor	
- amenities	45f.
- area	69
- footway	50
- seating	73
out-patient service facilities	87
parking space	49ff.
passage	95f.
passing place	43, 7
path	42ff., 91f.
patio	69
pedestrian	
- circulation area	70
- crossing	48f.
performance concept	19
Pestalozzi	12
place of accommodation	62, 73
place of assembly	56f., 73, 93ff.
places of assembly legislation	94
Places of Work Act	21
Places of Work Directives	21
planting	99
playground	45f.
playing apparatus	70
poor laws	14
population pyramid	29
positive contrast	38
principle of two senses	17, 37, 40, 92
priority level	37

Appendix

private bathroom	67f.
protection against scalding	58
providing information	73
public building	93 ff.
public space	50, 91ff., 105
public transport	47f., 70
publicly funded housing development	11
pull cord switch	59
radio frequency identification chips (RFID chips)	45
railway platform	47
ramp	51ff., 71, 81, 92ff., 102
ratio of pensioners to those of working age	31
ratio of persons of working age to those below the age of 20	31
reach	96f.
reach height	66
reanimation of urban centres	95
red-green colour blindness	38
reflexology path	46
repatriation of prisoners of war	15
residential home	13f., 22, 87
restaurant	73
revolving door	51
Rights of Persons with Disabilities	41
riser	53f.
room temperature	73
routing	99
sales display	61f., 73
sanitary	
- area	82
- appliance	78f.
- facility	57ff., 67f., 72f.
school	75
secondary footway	70
securing of building sites	70
seeing	37f.
senses	105
sensorial impairments	9, 35f.
sensorial issues	36
service	61f.
severe disability	17
sheltered housing with services	87
sheltered housing	13, 86ff.
- attached to residential home	13
- individual	13
- integrated	13
shopping	95f.
shower	59, 67f.
- cubicle	72
- tip-up seat	59f.
side-hung door	51
sign language interpreter	56
signs	72
soap dispenser	72
Social Code	20
spandrel panel	73
sports facilities	60f., 73
stair width	53
staircase	53, 84
stairs	45, 52ff., 64, 84, 92, 102
standardisation	17
standards	20
stepless accessibility	50f., 73
street	42ff., 105
subsistence welfare	15
support and grab rails	67, 72
support rail	67
supportive employment	97
surface characteristics	43f.
swimming pool	61
tactile guidance system	23, 48
tactile information	73
technical construction regulations adopted in building legislation	18f.
terrace house	82f.
therapy pool	61
ticket window	61f., 73
toilet	67f., 72
- facilities	57f.
- approachable from either side	58
- roll holder	72
- flushing	72
touching	99
touchscreen	40
traffic signals	48, 91
training area	45f.
transferring from wheelchair to toilet	67
transverse fall	50
tread	38, 53f.
universal design	15ff., 40
usability	41, 82, 97
vertical access	52ff.
visual information	73
waiting area	41
warning	73
wash-basin	58, 68, 72
waste receptacle	72
WC	72
welfare for cripples	14f.
wheelchair accessibility	17
wheelchair parking space	72f.
white cane	45
WHO	9
window	69
workplace	55ff., 96f.

Appendix

Picture credits

The authors and publishers would like to express their sincere gratitude to all those who have assisted in the production of this book, be it through providing photos or artwork or granting permission to reproduce their documents or providing other information. All the drawings in this book were specially commissioned. Photographs not specifically credited were supplied from the archives of the architects or the magazine DETAIL. Despite intensive endeavours we were unable to establish copyright ownership in just a few cases; however, copyright is assured. Please notify us accordingly in such instances.

Cover centre, page 53:
Klaus Neumann, Munich

Cover bottom, page 93 bottom:
Aldo Amoretti, San Remo

page 6, 104, 105:
Marc Wetli, Zurich

page 8:
Christina Merkan, Berlin

page 12:
Seeger-Ullmann Architekten, Munich

page 14:
Dreizung, Berlin

page 17:
Simone Rosenberg, Munich

page 19:
Levin Monsigny Landschaftsarchitekten, Berlin

page 23, 38, 42, 49, 51, 52, 61 top, 68 bottom, 82 bottom right, 106:
Oliver Heiss, Munich

page 34, 51 left:
Christian Schittich, Munich

page 35:
www.zeno.org, Zenodot Verlagsgesellschaft mbH

page 45:
www.archdaily.com

page 46, 76, 77 left:
Adam Mork, Copenhagen

page 54:
Hertha Hurnaus, Vienna

page 57 left, centre:
KEUCO GmbH & Co. KG

page 59:
illbruck Sanitärtechnik GmbH, Bad Wildungen

page 62 centre, 69:
Michael Heinrich, Munich

page 62 right:
Klaus Frahm/arturimages

page 64:
Johann Ebe, Munich

page 66, 67:
Stefan Müller-Naumann/arturimages

page 68 top:
www.bette.de

page 74:
Claudia Fuchs, Munich

page 75 top:
Jan Bitter, Berlin

page 77 centre:
Bodensteiner Fest Architekten Stadtplaner, Munich

page 78, 79 left:
Florian Holzherr, Munich

page 79 centre:
Tom Früchtl, Munich

page 80:
Hans Werlemann/Hectic Pictures, Rotterdam

page 81 left:
Andreas J. Focke, Munich

page 81 right:
Christine Dempf, Munich

page 82 top:
Rainer Retzlaff, Waltenhofen

page 83 top:
Jesús Granada, Seville

page 84:
Walter Mair, Zurich

page 85:
Bernd Borchardt, Berlin

page 87:
Rathscheck Schiefer, Mayen-Katzenberg

page 88, 89:
Zooey Braun Fotografie, Stuttgart

page 90 top centre, right:
Ronald Grunert-Held, Veitshöchheim

page 90 bottom:
Oliver Voitl, Munich

page 92 top left:
Claas Dreppenstedt, Berlin

page 92 bottom:
Erik Berg/Den Norske Opera & Ballett

page 94 left:
Martin Bosch, Bayerische Schlösserverwaltung

page 94 right, 97:
Stefan Müller-Naumann, Munich

page 94 bottom:
Ivan Nemec, Prag

page 95 top right:
www.fotolia.com

page 95 bottom:
www.pulte-fabrik.de

page 95 top left:
Frank Kaltenbach, Munich

page 96:
Dominique Perrault, Paris

page 98–100:
Yves André, St. Aubin-Sauges (CH)

page 101–103:
Stefan Müller, Berlin

Full-page plates

page 6:
"City Lounge", St. Gallen (CH), 2005; art: Pipilotti Rist; architect: Carlos Martinez, Widnau

page 8:
Lift call button in Berlin-Charlottenburg

page 34:
BMW Museum, Munich (D), 1973;
architect: Karl Schwanzer, Vienna
Architects, 2008 conversion: Atelier Bruckner, Stuttgart

page 74:
Opera House, Oslo (N), 2008; architects: Snøhetta, Oslo

page 106:
Braille writing